622
B211h

100777

622        Ballantyne, Verne
B211h                    100777
How and where to
find gold

DEMCO

# HOW AND WHERE TO FIND GOLD

## Secrets of the '49ers

# HOW AND WHERE TO FIND GOLD

## Secrets of the '49ers

**Verne Ballantyne**

ARCO PUBLISHING COMPANY, INC.
219 Park Avenue South, New York, N.Y. 10003

Published by Arco Publishing Company, Inc.
219 Park Avenue South, New York, N.Y. 10003

Library of Congress Catalog Card Number 75–18878
ISBN 0-668-03859-4


Printed in the United States of America

And the gold of that land
is good: there is bdellium
and the onyx stone.

*Genesis 2:12*

# Acknowledgments

My good friend and associate, Bill Little, was the important link between me and the old time prospectors and miners. The clues given him by the early prospectors are the basis for this book. The personal incidents related are his.

Terry Groth, a local young man in the sign business, caught the spirit of the text and by his creative talent produced the artwork which has added so much to the book.

The Montana Bureau of Mines graciously gave permission to use their working drawings of the sluice box and the cradle. Their evaluation of the pan as a prospector's tool and as a device for actual gold recovery was most helpful.

The California Division of Mines and Geology gave valuable help by authorizing the reproduction of a typical gold diving operation from their April 1972 issue of *California Geology*.

Robert M. Ferguson, of the U. S. Forest Service was most helpful in providing material and guidance in the section relating to Federal Regulations.

Nancy St. John gave invaluable help in editing the manuscript.

# Contents

# HOW AND WHERE TO FIND GOLD

## Secrets of the '49ers

# Preface

There is still plenty of gold to be found in the West! Contrary to what many people think, all the gold in the United States was *not* mined out by the old '49ers or by the Depression Miners of the 1930's. This may have been true of the famed gold bearing creeks of the Mother Lode Country of northern California, but many lesser known areas were not worked as extensively. There are miles of buried gold placers in ancient streambeds in the High Sierra Mountains of California and other places that have never been touched. Many geologists believe that there is many times more gold in the ground than has ever been mined!

There is gold to be found in many of the tributaries of the Snake, Columbia and Salmon Rivers in Washington, Oregon, and Idaho. The Bitter Root, the Clark's Fork, and the North and South Forks of the Flathead River in Montana are all part of that great river system. The Kootenai, a little known but scenic and dashing river that loops down from the new gold fields of British Columbia into northwestern Montana and the Panhandle Country of Idaho, also offers good possibilities.

The High Country of south central Montana is an empire of its own. It is traversed by a network of tributaries of the Madison, Jefferson, and Gallatin Rivers which meet near Three Forks, Montana to form the mighty Missouri. The old gold mining towns of Virginia City, Alder, and Nevada City on Alder Creek, to name only a few, testify to the presence of gold in the early days, and who is to say it was all mined out?

13

Also, the countless mountain streams in the Rocky Mountains of Western Canada, the Northwestern Territories, and Alaska are all part of the Gold Belt. Much of this is virgin country waiting for modern prospectors to seek their fortunes.

Consider this: gold mining did not die out because there was no more gold! The demand for labor and material during two World Wars put economic pressure on gold mining. The wages offered to boost industrial output pulled men from the gold fields, and then Government restrictions clamped down on non-strategic gold mining.

Following World War I, the fixed price of gold at $35 an ounce could not cover the cost of production. It was impossible for most of the hard rock mines to reopen, nor was there any incentive for placer miners to return to their diggings. Gold mining could not compete with the higher wages paid in other segments of the economy.

Between 1972 and 1976, the official price offered by the Treasury of the United States of $35 (and later $42) an ounce was challenged by a free world market price of $150 to $200 an ounce. Further increases are likely. Consequently, a new interest in gold mining is growing proportionately.

In 1974, deteriorating confidence in so-called lawful (paper) money, with no backing in gold or silver and very little purchasing power, prompted the 93rd Congress to repeal the infamous Gold Reserve Act of 1934 which had made private ownership of gold illegal. Thus a right which had been denied United States citizens for over 40 years was again restored.

The effect of this Act on gold mining in the United States, coupled with the phenomenal rise in the price of gold on world markets, can only be guessed. Could it be that the Gold Rush of the 1970's is just beginning?

Verne H. Ballantyne
Bozeman, Montana

# Introduction

In American folklore, a bewhiskered old man leading a burro symbolizes the search for gold. The prospectors who found the prized yellow metal in the deserts and mountains of the West did so with pick, shovel, gold pan, and sweat. They survived on bacon, beans, coffee, and grim determination. Hard, patient work and trial and error developed their knowledge of how and where to find gold. This precious knowledge, born of experience and only shared with other old timers, produced a reservoir of secrets not found in any textbook. The campfires where tall tales and true adventures were told were the prospector's trade schools. Wherever gold prospectors met, the talk was about how and where to find gold. Although the old gold mining days are history, the secrets of the old prospectors have lived on. This book will tell you about these colorful men and the mining skills they used in the days when gold mining was a way of life and a crucial economic venture.

You may be wondering how I came to have the gold-finding knowledge of the old timers. That is an interesting story in itself. In the mid-1900's, some of the old timers were still around and actively seeking gold. As a young man just out of the Air Force, I was attracted to the old-time prospectors and the sourdough. I have sat by the campfire many a night and re-lived through these men their experiences on the trail of gold, and I have listened to their theories about how Nature made gold and hid it away for men to find in due time and season.

These men were eager to pass on their secrets. They let me work

*Gold mines were not closed because there was no more gold! They were closed because the Government's fixed price of $35 per ounce would not permit the miners to cover the cost of production.*

*Prospector's seminar.*

with them on their claims and taught me how to work a digging. I became an amateur member of their circle and eventually staked my first claim. That was over twenty-five years ago. The search for the yellow metal has been a prime interest in my life, along with an interest in related mining operations. The wisdom of the early prospectors has been enhanced by modern technology. The burro or pack horse has been replaced by the four wheel drive pickup and camper. But nothing can replace the gold pan and pick, even though a gas-engine powered suction pump and portable sluice box come in handy. The modern miner still has a lot to learn from the old prospectors about where to look for gold and how to find and recover it. In this book, we have combined the best of the old secrets with newer methods and equipment.

My hat is off to you, partner, and may your prospecting days be many and your rewards be great.

William A. Little
LaPorte, California
1976

# Chapter I
# Where to Look for Gold

Until the old '49ers found gold in California, no one except the Indians and the Spaniards knew it was there. It is common knowledge now that there is gold in Arizona, Nevada, Montana, South Dakota, Canada, Alaska, and in some of the Southeastern states. That doesn't mean you can find gold just anywhere in those states, or even in specific areas of those states, but it does help to know that there are certain parts of the country where gold definitely exists. (See page 20.) The first rule to guide you in your search for gold is:

Look in places where Gold has been found in the past.

You will probably do your looking for gold close to where you live, as this is likely to be a weekend or part-time venture. Write to the Bureau of Mines in your state and request information on the location of known gold fields. If you are seriously considering looking for gold on a full time basis, write to the United States Bureau of Mines in Washington, D. C., and request maps and information on known gold producing areas in the United States and Canada. You might consider leaving home in this case—but think twice before you do. Prospecting is a highly speculative activity even for an experienced person. It carries no guarantee of providing a living.

## Placer Gold

Placer is a word of Spanish origin defined as "an alluvial or glacial

19

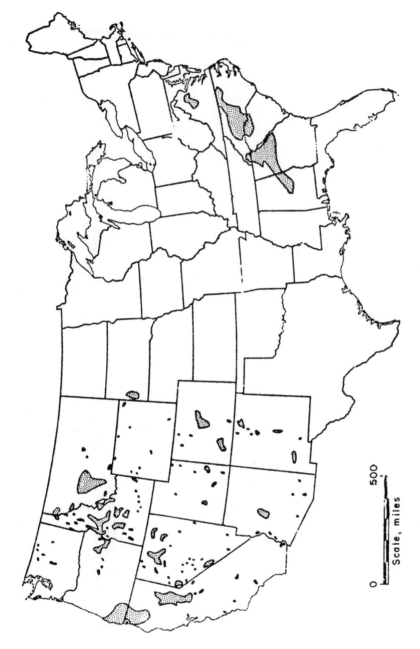

Scale, miles

0       500

*Gold placer areas of the United States (Courtesy U.S. Department of the Interior, Bureau of Mines, Information Circular, How to Mine and Prospect for Placer Gold).*

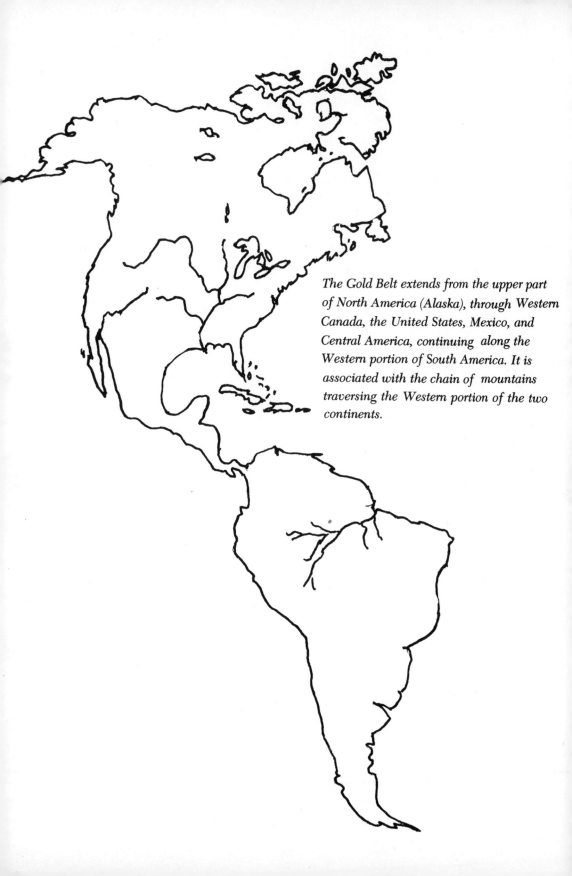

The Gold Belt extends from the upper part of North America (Alaska), through Western Canada, the United States, Mexico, and Central America, continuing along the Western portion of South America. It is associated with the chain of mountains traversing the Western portion of the two continents.

deposit containing particles of gold or other valuable mineral."
Undoubtedly, the first discoveries in North America by the Indians of
the attractive yellow metal were from placer deposits. These were
later worked by the Spanish.

The historic discovery in California at Sutter's Mill in 1849, in what
is now Amador County, was a placer discovery and most of the early
gold rush was for the gold found in the rivers and streams of
California, Southern Oregon, and the other western states.

### Quartz—A Source of Gold

The old timers agreed that the free gold found in the streambeds
and bars in the form of nuggets, flakes, and dust had to come from an
original source, and this source was usually quartz. For some reason,
unknown to the old timers or to modern geologists, gold is almost
always found in or very near quartz rock.

Weathering, upheavals, and erosion broke up the quartz and
gradually released the gold from the veins or streaks in the rock. The
released gold was then subjected to the forces of erosion, and the
grinding or polishing action of rock and water. Small pieces of gold
were literally ground to dust, and larger pieces were reduced to small,
rounded nuggets. This was the source or origin of the free gold known
as placer gold.

### Gold Is Heavy

Aside from its yellow color, the most distinguishing characteristic of
gold is its weight. It is 19.7 times heavier than water and at least seven
times heavier than any rock with which it is found. The weight of gold
allows us to separate it from other mineral material. This factor is basic
in most of the equipment used in the recovery of gold, as we shall see
later.

### Hard Rock Gold

The gold found in place in the quartz rock is called vein or hard
rock gold and its recovery requires an entirely different procedure

than the methods used in recovering placer gold, which will be discussed later.

## Bedrock

Bedrock is the solid rock mass underlying the superficial formations and soil. Knowing about bedrock is important to us because a large proportion of placer gold is found on top of bedrock. Because of its extreme weight, gold usually washes to the bottom of the surface material. Not all gold finds its way to bedrock but that is where the larger nuggets and flakes are usually found. The illustration on page 24 shows likely locations on the bedrock where gold may be lodged.

## Potholes, Riffles, and Rapids

The inside bend of the creek where large boulders or outcroppings are present is a likely place to find gold. Gold tends to lodge behind or on the downstream side of these boulders, in pockets or potholes formed by the current of the stream. Collect your gravel samples at varying distances from the boulder, as close to the bedrock as possible. Another promising place is just above any natural shelf which crosses the stream bed. Shelves are indicated by riffles in the stream where the water slows to form a pool where gold may collect. The heavier gold will settle out first and sink to the bottom of the streambed. Natural breakwaters created by the bank or by a fallen log may also provide places where the running stream will drop its gold. These areas are called creek placers to distinguish them from other types and places of gold accumulation, and they are considered to be the most important location for placer gold.

## Bench or High Bar Placers

Another important source of gold is the area immediately above the present creek bed. Along these banks there may be gravel deposits or bars on top of the bedrock where the gold has been deposited along what was formerly the wider creek bed.

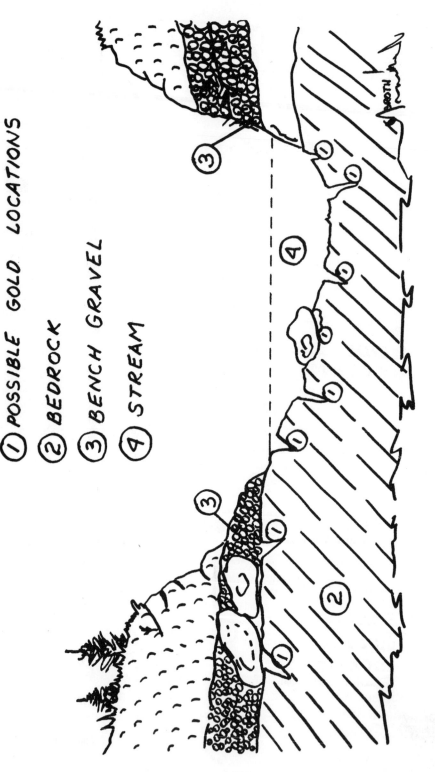

POSSIBLE GOLD LOCATIONS

① POSSIBLE GOLD LOCATIONS
② BEDROCK
③ BENCH GRAVEL
④ STREAM

*Cross-section of a stream bed.*

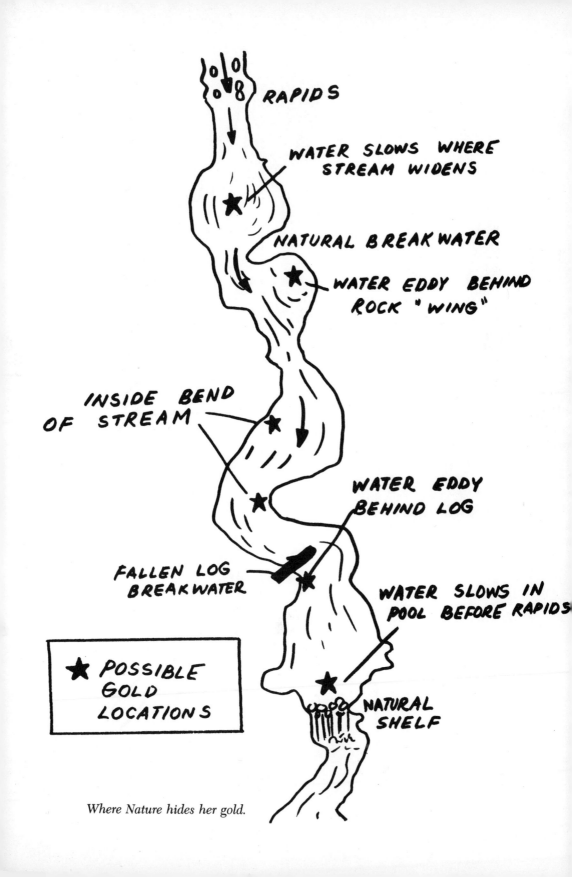

RAPIDS

WATER SLOWS WHERE STREAM WIDENS

NATURAL BREAKWATER

WATER EDDY BEHIND ROCK "WING"

INSIDE BEND OF STREAM

WATER EDDY BEHIND LOG

FALLEN LOG BREAKWATER

WATER SLOWS IN POOL BEFORE RAPIDS

★ POSSIBLE GOLD LOCATIONS

NATURAL SHELF

*Where Nature hides her gold.*

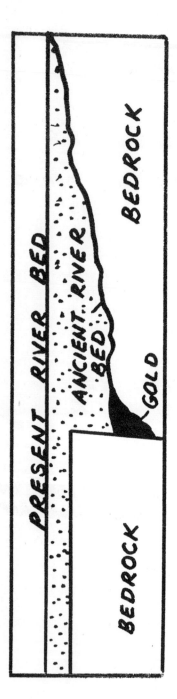

Cross-section of a gold-bearing desert stream valley (Manhattan, Nevada), showing the results of several periods of stream deposition from the oldest (1) to the youngest (6). (After Ferguson, U.S. Geological Survey Bull. 723.)

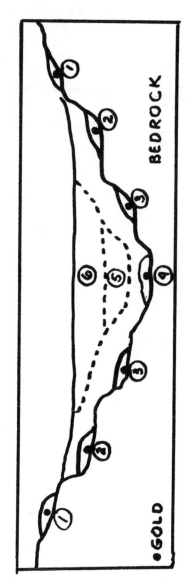

Ideal cross-section of a river in the Sierra Nevada, the bed of which has suffered down-faulting on the upstream side, causing gravels, sand, and silt to accumulate in the pocket thus formed. (Placer Mining For Gold in California, Div. of Mines, State of California, Bull. 135, Sec. II, Jenkins.)

## Crevice Mining

The beginning prospector can start his search for gold in several ways and with varying amounts of equipment. One approach that takes a minimum of equipment and will get you started in this most interesting and challenging activity is to do some crevice mining. Crevice mining is a specialized type of placer mining. The name comes from the fact that one of the most likely places to find gold in a mountain stream is in the crevices of the bedrock over which the stream flows. Gold has accumulated in these cracks and crevices over a long period of time.

The old timers found that gold was located in the cracks and crevices in the bedrock not only in the present streambed but also higher up the bank and in exposed bedrock that the stream had reached during spring runoff. Gold was hidden in dry streambeds as well. Some of the early miners made a special search of these cracks and crevices. They were called crevice men or crevice miners. Crevice men who didn't bother to stake claims and ranged up and down the creek working on any ground not staked were called snipers.

### *Tools Required*

Crevice work is probably the easiest way to start looking for gold. To be a crevice miner, you need specialized tools to get down into those cracks in the bedrock. I'm not going to send you to the store to get a lot of expensive equipment. You probably have some of the tools you'll need at home. Here is a list of the crevice man's tools:

1. Large tablespoon.
2. Long handled teaspoon (tall glass variety).
3. Old screwdriver, bent two to three inches from the end of the blade. Makes an ideal digger.
4. Tweezers. These are for picking up flakes or small nuggets out of the sand in your pan.
5. Small glass jar with a tight fitting screw cap with a wide mouth. Put your gold in this.

6. Small magnet. This is useful in sorting out the heavy magnetic pieces of black sand after your panned sample has dried.
7. Small shovel with short or long handle, whichever you prefer.
8. Gardener's hand scoop.
9. Miner's hand pick.
10. Magnifying glass. This will add interest and make identification easier. The folding, double lens type is preferred.
11. Flashlight.
12. Galvanized metal bucket. It has many uses and is handy for carrying your small items to the site of your diggings.
13. Gold pan.

The last and absolutely indispensable item is a gold pan. Some of the old timers may have used a frying pan in a pinch, but I am sure they got a regular gold pan at their first opportunity. It is designed and shaped for panning gold. Several sizes are available, but the size generally referred to as standard is 16 inches in diameter at the top and 10 inches at the bottom, with a depth of $2\frac{1}{2}$ inches. I recommend the half size pan which has a top diameter of 12 inches, a $7\frac{1}{2}$ inch bottom, and a depth of 2 inches. Level full, this pan weighs 9 pounds—as compared to 20 for the 16 inch pan. It is easier to carry and easier to use in a small stream or tub. Its lighter weight when filled is less tiring. An experienced panner can wash two of these half sized pans of gravel in less time than it takes to wash one standard size pan.

A new pan is usually coated with some type of oil, grease, or other rust preventive. This must be removed before using the pan, as even the slightest trace of any oily substance (even body oils which may accumulate from your hands in normal use) will cause fine gold particles and dust to ride out over the edge when panning. You must clean your pan before using it. The surest and easiest way to clean your pan of this oil is to pass it back and forth over a gas stove burner or similar flame until the metal turns blue. Even though this is called burning or blueing the pan, care must be taken not to heat it unevenly or excessively. Excessive heat under one spot can warp your pan and make it useless. Blueing the pan not only does away with the grease, but also provides a dark blue color which is an ideal contrast to fine gold particles.

*Crevice miner.*

Crevice miner's tools.

Gold pan

Pail

Bent file

Tablespoon

Long-handled teaspoon

Magnifying glass

Tweezers

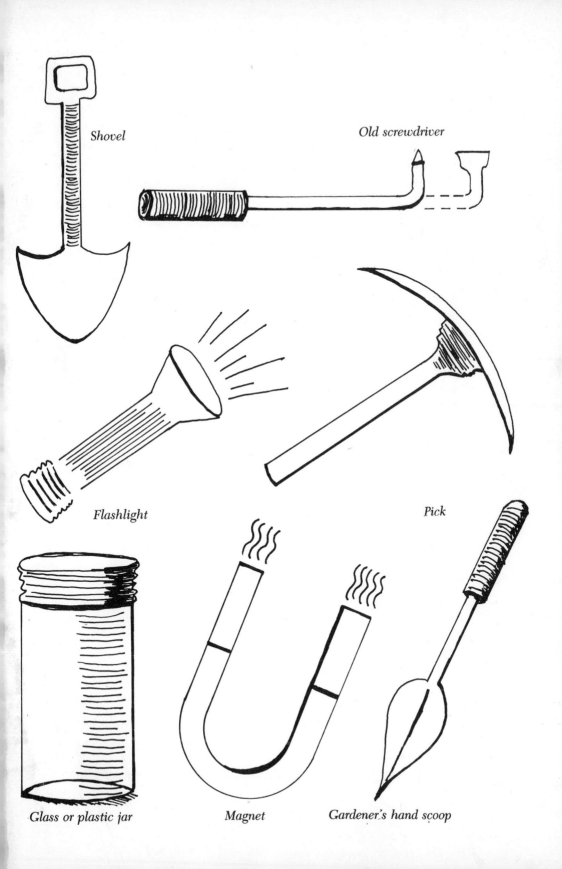

Shovel

Old screwdriver

Flashlight

Pick

Glass or plastic jar

Magnet

Gardener's hand scoop

## Mining the Crevice

Mining the cracks, crevices, and fissures in the bedrock is our present objective. Pick out your crevice. It is probably filled with silt, sand, small gravel, and vegetable matter. Using the bent screwdriver, loosen this material so that you can scoop it out of the crevice with your spoon and dump it in your gold pan. Dig all the way to the bottom of the crevice. Remember that the gold, answering the call of gravity, has probably been working its way to the bottom of that crevice for a long time. You may have to enlarge the crevice or open it up a little with your pick. Go ahead and dig down into the crevice! This is the way to learn on your own.

Fill your pan two-thirds full. Include the plants growing in the crevice. Their roots will have entered the very narrowest bottom part of the crack and the tiny root hairs may have grasped a flake or small particle of gold. If so, it will be released when you shake the roots around in the water in your pan.

Now you're ready to pan your sample!

# Chapter II
# Panning for Gold

You have a pan and are ready to take a first sample. Follow these steps until they become natural and automatic:

## 1. Preparation

Having filled the pan two-thirds full of sample material from the crevice, choose a shallow place along the bank of the creek that is a little deeper than your pan. The stream should be moving fast enough to keep the water clear, but slow enough so that it won't wash any of your sample away. (A tub of water will do if you aren't near a creek.) Carefully submerge your pan until it is resting on the bottom of the creek. When the material is thoroughly wet, work and stir the mixture around with your hands until any lumps of dirt are disintegrated. Break any hard lumps between your thumb and fingers. Dirt and sand adhering to any plant roots should be washed into the pan. Then throw out the plant material. Any clay present must be stirred up until it is entirely dissolved and washed away. The large gravel and rock can be picked out and discarded.

## 2. Suspension and Stratification

Next, grasp the still submerged, level pan by its opposite sides and perform a vigorous left-right, left-right motion. This, plus a clockwise,

*One of the last of the old time prospectors demonstrates the fine art of panning for gold. Taken in the summer of 1955.*

counter-clockwise motion, should keep the contents loose but not agitated. This permits the separation of heavy materials from lighter ones. The heavier minerals will sink to the bottom while the lighter gravel and rock material will be displaced upward. Most of the larger gravel can then be raked over the edge of the pan with your fingers.

## 3. Washing

Washing is accomplished by slightly raising the near side of the pan above the water, leaving the far side submerged. Switch from the right-left, right-left motion to a circular, swirling one. This will gradually work the lighter material over the submerged side of the pan into the water. The edge of the pan can be raised or lowered to regulate the washing. A side-to-side motion is also useful at this point, used alternately with the circular action.

## 4. Cleaning

The objective of this step is to wash away the remaining sand and pebbles from the heavier material on the bottom of the pan. The pan is jerked vigorously from side to side to loosen the material and cause further stratification. At the same time, the pan is tipped forward gradually, until the sand bed is flush with the pan's edge. Stop briefly at this point to allow the bed to settle.

Your pan is now at an angle, with the lower edge barely under water. Maintaining this angle, dip the pan vertically into the water four or five times, so that the bed is slightly covered. As you raise the pan, you will see lighter particles wash out over the edge into the water. When this lighter material is washed away, the pan is again partially filled with water, and the bed loosened and restratified by repeating the vigorous side-to-side motion. Repeat the dipping, washing, and shaking until only the heavy mineral concentrate remains in the pan. This may seem awkward or tedious at first, but you will soon develop your own variations on these basic techniques.

## 5. Inspection

The panning operation is complete when the original sample of material has been reduced to a small quantity of black sand and minerals. Inspection of the sample is made by placing a small amount of water in the pan and swirling it gently, so that the water moves the lighter particles ahead of the heavier ones. The gold is brought into view at the end of the wash of material, if you have been working in gold country.

## 6. Recovery

Larger pieces of gold can be picked out with your tweezers and dropped into your glass jar. Smaller pieces and flakes will adhere to the end of a wooden match long enough to be put into the jar. If there is a considerable amount of fine, black sand, the sample should be dried and the sand removed by placing it on a dry pan, or a stiff piece of paper or cardboard. When the sand is dry, blow gently across the surface while tapping the pan or paper. The sand will quickly blow away from the gold and any heavy magnetic material. Your pocket magnet can then pick up the magnetic material and the gold can be scooped up with a stiff piece of paper and poured into the flask. If the magnet is wrapped in plastic, the material on it can be discarded easily when the magnet is removed from the plastic.

*1. Preparation—Mix and stir to separate all the parts of the sample.*

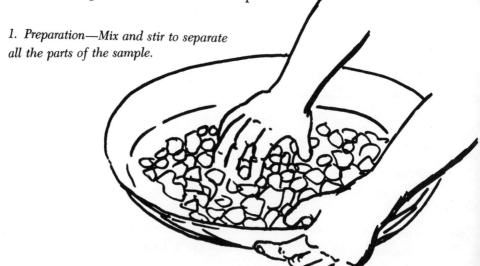

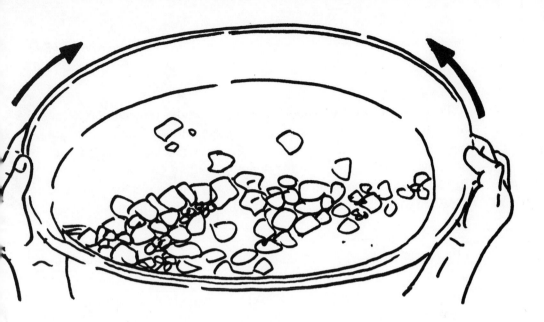

2. *Suspension and stratification—Give the pan sharp jerks clockwise and counter-clockwise. Rake large waste over the side of the pan.*

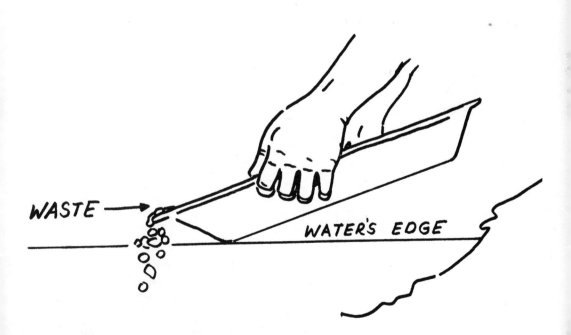

WASTE →

WATER'S EDGE

3. *Washing—Tip the pan and give the pan a circular motion to dispose of waste material.*

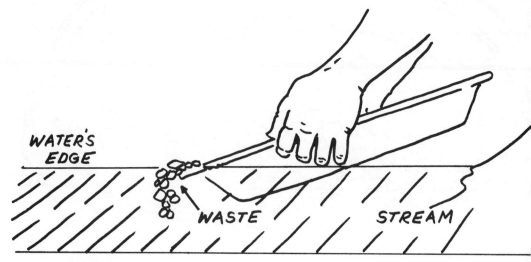

WATER'S EDGE

WASTE

STREAM

4. *Cleaning—Dip the pan several times up and down, and vigorously from side to side.*

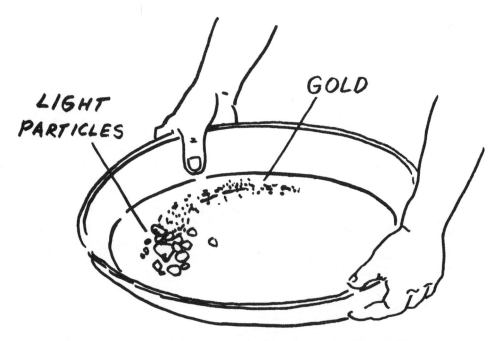

LIGHT PARTICLES

GOLD

5. *Inspection and Recovery—Swirl contents around with a small amount of water to move light particles ahead and expose colors. Remove gold with tweezers.*

## Mercury

Mercury is often used to separate the fine gold particles and dust from the black sand. When the sand and gold mixture is thoroughly dry, pour a little mercury into the pan and roll it around until it picks up all the gold. To recover most of the free mercury, pour the gold-mercury amalgam onto a piece of chamois skin or fine canvas, then fold the four corners over to form a pocket containing the amalgam and squeeze as much of the mercury as possible through the skin or canvas into a dish. The remaining ball of amalgam can be reduced to its separate components with this trick used by the old timers. Place the amalgam in a hole scooped out of the center of a raw potato. Then place a tin plate over the hole, turn all of it over, and heat it on a stove until the potato is partly cooked. The gold will be found on the plate in the form of a sponge. The heat will vaporize the mercury (as it has a very low boiling point), and it will condense and remain in the cooler flesh of the potato. The mercury can be recovered by placing the potato in a dish of water. The mercury will sink to the bottom. *EXTREME CAUTION must be taken so that the mercury vapor does not escape into the air. MERCURY VAPOR IS VERY POISONOUS.*

## Grizzley Pan

A trick that will speed up the panning operation immensely is the use of a sieve called a grizzley pan. It is made by drilling a number of quarter inch holes in the bottom of a pan the same shape and size as the one used for panning. Place the grizzley inside the regular pan, fill it with your pay gravel, and submerge it in the water in the usual manner. When the mixture is well soaked, lift the grizzley slightly and rotate it sharply back and forth under water until all the material less than a quarter inch in diameter has passed through into the regular pan. The larger material can be checked quickly for any large nuggets and then can be discarded. The fine material in the regular pan is washed in the usual way.

## Safety Pan

Occasionally, even an experienced panner will accidentally spill his carefully washed sample into the creek. To prevent the use of profanity and to save what might be a valuable sample, it is a good practice to direct the pan tailings into a second pan which is called a "safety pan." You can decide for yourself if the use of the safety pan is worth the added bother.

## Practice Panning

Successful panning is a skill that is acquired largely by practice and experience after a certain amount of instruction. We suggest that you practice at home before you head off to seek gold. Conditions very similar to those encountered in placer mining can be simulated by using ordinary sand and gravel, your gold pan, a tub of water, and fifty or so buckshot or very small fishing sinkers. You can even cut some of the shot up to simulate fine particles of gold. This isn't very exciting or romantic, but it *is* practical. With your gold pan and these materials, you can follow the instructions and test your proficiency. Fill your gold pan two-thirds full of sand and gravel and then add the fifty buckshot. If you get to the bottom of your pan and find only half of the shot left you are doing well for a beginner, but you can do better. Suppose those other 25 buckshot had been gold nuggets that you washed into the creek! Practice until you find 45 to 48 shot left in your pan. Then you will know that you are good enough to get the most from your samples.

## Fool's Gold

One particular bit of mining folklore had to do with "fool's gold," the common name given to iron pyrite. Chemically speaking, iron pyrite is a combination of iron and sulfur. It is yellow and it frequently appeared in the bottom of the gold pan. It was called fool's gold because inexperienced miners mistook it for the real thing. It can be easily distinguished from gold, however, as it is easily crushed. Gold is highly malleable; it will not crush or break up. Iron pyrite is deceptive. Oddly, gold is often found near it, sometimes in the same rock

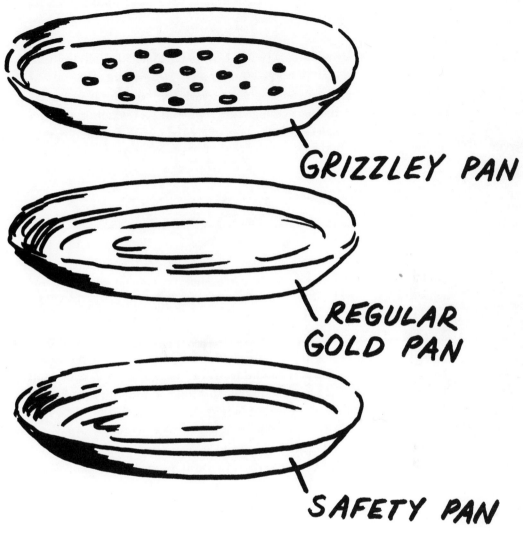

GRIZZLEY PAN

REGULAR GOLD PAN

SAFETY PAN

*The three pans.*

formation. For this reason, the old prospectors did not ignore fool's gold.

### Evaluation of Panning

To put this whole matter of panning in perspective, a few observations can be made. Panning is an indispensable way for the placer miner to analyze the gold content of a prospective diggings that may then be worked by other, more rapid and productive means such

*Practice panning.*

as the cradle or sluice box. The gold pan alone has recovered a good amount of gold for old-time miners, Depression miners, hobby and Sunday afternoon miners and other amateurs, but it does have its limitations from an economic point of view.

## The Economics of Panning

According to the estimates of the Montana Bureau of Mines and Geology in their Miscellaneous Contribution No. 13, an experienced man working with a standard size gold pan will do well to pan sixty pans in a ten hour day. At twenty pounds per pan, he will handle 1200 pounds of gravel. At this rate, it will take him three days to pan one cubic yard. If the gravel produces $15 worth of gold per cubic yard, which is fairly good gravel at a price of even $100 per ounce of gold, it is easy to see that it would take an entire day to recover five dollars worth of gold. The average for the Depression miner looking for gold selling for $35.00 per ounce was more like one dollar per day. Consequently, panning gold under average conditions is not a bonanza. It is a prospecting tool and should not be considered a good

*Frank Riley, foreman of the Feather's Fork deep channel placer mine when it was closed down in 1934, washes some tailings from the old dump in August 1974.*

method of serious or commercial gold recovery, except under exceptional circumstances.

## Panning for Fun

Of course, not everything is done strictly for economic reward. From the point of view of a hobbyist, weekend camper, or prospector, the reward of panning for placer gold may come largely from being in a natural and perhaps scenic environment where the air is fresh and the skies are blue. Any additional reward in the form of gold recovered is an extra benefit.

# Chapter III
# Recovery of Placer Gold

Panning has a romantic and rustic appeal, but there are other, more productive, simple methods of gold recovery. We'll discuss a couple of these.

## The Cradle or Rocker

The cradle or rocker was widely used by the early miners and it is still a valuable tool for recovering placer gold from gravel. The cradle is light-weight, easily and cheaply constructed, durable, readily transported, and can be operated by only one or two people. It will handle three to five cubic yards of gravel in a ten hour day and it will save all but the very smallest of gold particles. Details of its construction are found on page 46. Its operation is simple. A stream of water equal to the flow from a garden hose is fed into a screen-bottomed hopper containing a load of gravel. At the same time, the device is vigorously rocked back and forth. This action washes and sorts the gravel and sand containing the gold. The larger gravel is thrown out of the hopper by hand to allow for the addition of new material. The sand and gold drop through the screened bottom of the hopper into a riffled trough, the slope of which is regulated by raising or lowering either end of the bed plate. At the end of a day's work, any gold that has accumulated in the riffles is obtained by removing the riffles and flushing the sand down the trough into a tub. The gold is recovered by panning this sand.

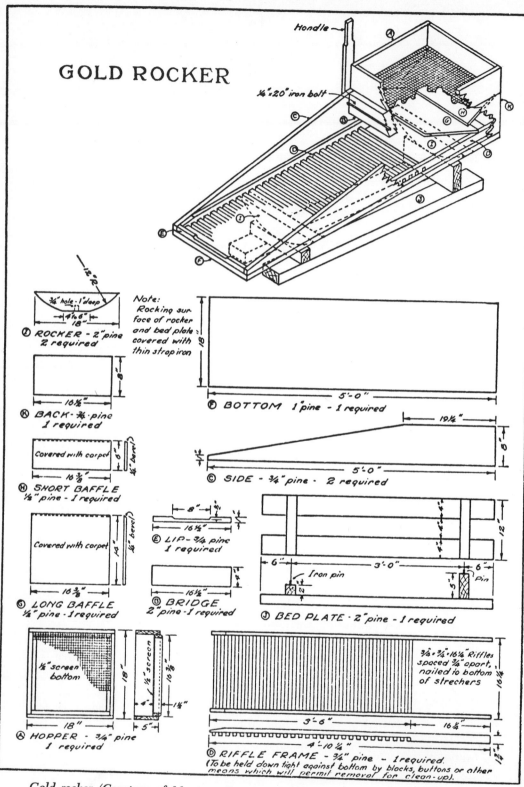

# GOLD ROCKER

Handle

1/4"-20" iron bolt

Note:
Rocking surface of rocker and bed plate covered with thin strap iron

**ⓛ ROCKER - 2" pine 2 required**
3/4" hole - 1" deep
4" to 6"
18"
12"-2"

**ⓚ BACK - 7/8" pine 1 required**
16 1/2"
8"

**ⓗ SHORT BAFFLE 1/2" pine - 1 required**
Covered with carpet
16 3/8"
6"
1/2" bevel

**ⓖ LONG BAFFLE 1/2" pine - 1 required**
Covered with carpet
16 3/8"
14"
1/2" bevel

**ⓔ LIP - 3/4 pine 1 required**
8"
16 1/2"

**ⓑ BRIDGE 2" pine - 1 required**
16 1/2"

**ⓕ BOTTOM 1" pine - 1 required**
18"
5'-0"

**ⓒ SIDE - 3/4" pine - 2 required**
19 1/4"
5'-0"
8"

**ⓙ BED PLATE - 2" pine - 1 required**
6"
3'-0"
6"
12"
Iron pin
Pin

**ⓐ HOPPER - 3/4" pine 1 required**
1/2" screen bottom
1/2" screen
18"
16 3/8"
18"
5"
1 1/4"

**ⓓ RIFFLE FRAME - 3/4" pine - 1 required.**
(To be held down tight against bottom by blocks, buttons or other means which will permit removal for clean-up).
3/8 x 3/4 x 16 1/4" Riffles spaced 3/8" apart, nailed to bottom of strechers
16 1/4"
3'-6"
16 1/4"
4'-10 1/2"

*Gold rocker (Courtesy of Montana Bureau of Mines and Geology, Memoir No. 5).*

It is important that the water and sand be forced to flow *over* the riffles. If the mixture flows under the riffles, it will carry the gold out the end of the trough. A piece of canvas, blanket, or other fabric can be put between the riffle frame and the bottom of the trough to prevent this, with the riffle frame wedged tightly against the bottom. The baffles are covered with carpet, burlap, or some similar material. They serve to spread and feed the gravel evenly down to the riffles.

### The Sluice Box

The sluice box is an even easier piece of equipment to make than the cradle and it is probably more widely used. (See the construction details illustrated on page 49.)

Ordinarily, each sluice box section is 12 feet long and 12 inches wide on the inside, with two sides and the bottom forming a trough. The sluice box is made of one or two inch material, depending on whether it is to be used as a portable unit or as a permanent one. The riffles are most important. They are held in place with a frame that holds them tightly against the bottom so that no water can pass under to carry the values out of the box. Square pieces of wood placed crossways in a frame are the most commonly used riffles, but other types frequently used include poles placed crossways or lengthways in the box. A layer of larger rock can even be used.

To keep the box from sagging, warping or otherwise losing its shape—which results in leaks and inefficient operation—the box must be properly supported and braced. (See page 49.) The slope of the box is very important and varies somewhat depending on the amount of water flowing into the box and the nature of the aggregate being washed. Six to twelve inches is an average gradient for a 12 foot box. First level the box with a carpenter's level; then raise or lower one end to get just the right amount of slope for your needs.

Although both the cradle and the sluice box are simple devices there are a number of details to consider, not only in their construction and operation, but also in the type and arrangement of the riffles. If you intend to build and operate either a cradle or a sluice box, it is recommended that you consult one or more of the references listed in the back of this book. One special source is Miscellaneous Contribution No. 13, published by the Montana Bureau of Mines and Geology in Butte, Montana: *Practical Guide For Prospectors and Small Mine Operators in Montana,* by Koehler S. Stout.

Both the sluice box and the cradle require a water flow, so it would be wise to confine your search for placer gold to creeks and rivers or to areas where water can easily be diverted or pumped to your washer. Old flumes and ditches along the mountain-sides of the West remind us of some miner's work to bring water to his diggings in an out-of-the-way dry spot.

## Mechanical Equipment

One device that the miners of the 1930's found to be of great value that the prospectors of an earlier day would have given their eye teeth to have owned is a small, portable, gas-engine-powered pump with a long flexible hose similar to a fire hose. With this, they could set up their sluice box near a rich bench or bar some distance from the creek or river and pump the water to their box. It saved them a great deal of work in transporting the pay gravel to a sluice box near the creek.

In this modern age of power equipment the placer miner can be in business on practically any scale, ranging from a small, portable pump, to a much larger operation using a bull-dozer, power shovels, and draglines to remove the overburden (the useless material above the pay gravel) and deliver the gold-containing gravel to a trommel or a large, mechanical washing plant. Of course, this type of operation requires a large amount of money; still, several hundred cubic yards of gravel can be washed in a day and a much lower grade of pay gravel can be handled than in a hand operation.

## The Floating Dredge

The huge, unsightly piles of gravel waste in the beds of many creeks in the western states are evidence of the use of enormous floating dredges to work the gold-bearing gravel back in the early days. These monsters inched their way up the creek bed, gobbling up the earth and discharging the gravel behind them. In their day, they were fairly efficient machines. One by one, however, they fell into disuse due to increased operating costs and the fixed price of gold. Some were

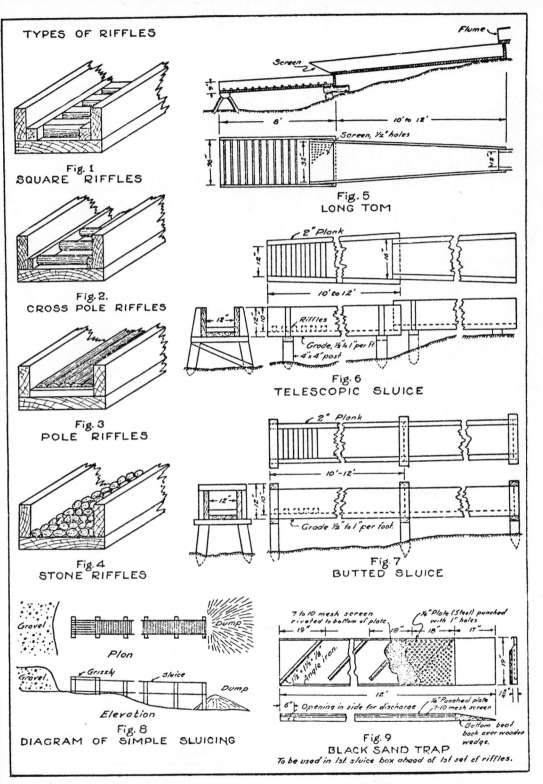

TYPES OF RIFFLES

Fig. 1
SQUARE RIFFLES

Fig. 2.
CROSS POLE RIFFLES

Fig. 3
POLE RIFFLES

Fig. 4
STONE RIFFLES

Fig. 5
LONG TOM

Fig. 6
TELESCOPIC SLUICE

Fig. 7
BUTTED SLUICE

Fig. 8
DIAGRAM OF SIMPLE SLUICING

Fig. 9
BLACK SAND TRAP
To be used in 1st. sluice box ahead of 1st set of riffles.

*Sluice box (Courtesy of Montana Bureau of Mines and Geology, Memoir No. 5).*

*Floating Yuba dredge. This dredge, located on Prickly Pear Creek near Jefferson City, Montana, between Helena and Boulder on U.S. Highway 91, was built 4¼ miles from its present site in 1938. It operated successfully until 1958, when it was closed down due to the reduction of gold content of the gravel. During its period of operation it produced more than 65,000 ounces of gold—worth two and a quarter million dollars at $35 per ounce.*

*The maximum capacity of the dredge is 8,000 cubic yards of gravel per 24 hour period. It is powered by electricity and with minor repairs it would be in operable condition. With the increased price and interest in gold it may again serve a useful purpose, probably in another country where ecological standards are not as high as in the United States.*

dismantled and shipped to other countries where operating costs were lower, but others have remained on the site of their last operation as ugly reminders of that period in the history of placer gold mining.

Typical of these huge floating dredges is one near the highway in a pond on a small creek near Jefferson City, Montana, between Helena and Boulder. A picture of this representative of these old dredges is found on page 50.

### The Modern Portable Floating Dredge

There is a modern, lightweight, portable version of the floating dredge which is both inexpensive and efficient. It combines scuba diving gear; a rubber raft or other platform; a light, efficient gasoline engine; suction equipment; and a portable sluice box. These dredges work on the same principle as a vacuum cleaner and can be used in ponds, rivers, or creeks.

Initial underwater exploration can be done with a snorkel, to avoid setting up the floating dredge if the location proves to be unfavorable. The objective of the floating dredge operator is the same as the crevice

*Modern floating dredge. Small floating dredge on Sucker Creek, which is a branch of the Illinois River near Caves Junction, Josephine County, Oregon, in August 1974.*

man, except that he is working underwater. In addition to suction equipment, he often uses crevice tools to work the deep and narrow cracks where the gold is most likely to be found. The vacuum hose of the suction dredge picks up and delivers the sand and gravel from the crevices and other traps to the upper end of the portable sluice box on the floating platform.

This method of mining placer gold was not even dreamed of by the old prospectors of the early days or miners in the Depression days of the 1930's. With equipment such as this, the modern gold seeker has an enormous advantage over the miners who have gone before him.

For anyone interested in more information on the floating dredge, one of the most complete sources is the June 1972 issue of *California Geology*, available from the California Division of Mines and Geology, P. O. Box 2980, Sacramento, California 95812; the price is 25¢.

## TYPICAL GOLD - DIVING OPERATION

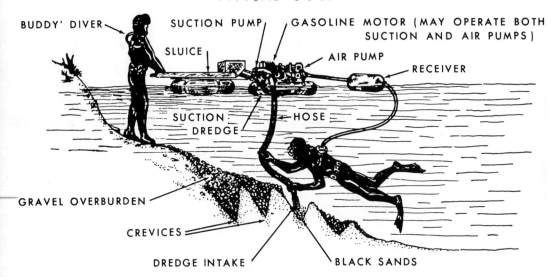

*In the typical gold diving operation, the gasoline-engine powered air compressor and suction pump can be located on shore or mounted on flooats so they can be towed by the diver as he works under water. (From "Diving for Gold" by William B. Clark, California Geology, April, 1972, California Division of Mines and Geology.)*

## The Floating Jet Dredge

The modern suction or jet dredge uses the Venturi tube principle. A motor driven pump on the surface feeds water under pressure to the head of the Venturi tube where suction is created by the stream of water passing swiftly into the larger tube. The created suction, or vacuum, in a flexible tube is directed by the operator to crevices and cracks and along the face of the bedrock of the stream, or pond. The intake from the suction is run into a sluice box where the riffles collect the gold as the gravel and water passes through to the discharge end of the sluice box.

This pipe-like device ranges from 4 to 8 feet in length and weighs up to 20 pounds. It is usually made of galvanized sheet metal and the curved or intake end is of stainless steel or other resistant material.

The jet dredge and the suction pump dredge both use "suction," or a partial vacuum principle, to suck up the gold bearing sand and smaller gravel from the crevices and other likely places on the streambed where gold may have accumulated.

Unlike the jet dredge, the suction pump dredge creates the partial vacuum by the use of a vacuum air pump driven by a small, portable gasoline engine which may also operate the air compression pump.

## Deep Channel or Buried Placers

Another type of placer in the gold country is an ancient streambed sometimes called a deep channel, or buried placer. If a creek or river changed its course due to a rock slide or some other natural event, the dry streambed containing gold-bearing gravel soon filled with weathered rock and debris. Even major earth changes may have occurred; some buried placers are now covered by clay and lava. Some of the most productive placer mines in the Mother Lode section of California were buried placers. These deep channel or buried streambed placers were recognized to be one of the most productive sources of placer gold when the mines were closed down during World War I. Many potentially valuable deep channel placers are still lying undiscovered, waiting for some knowledgeable and industrious prospectors to find them.

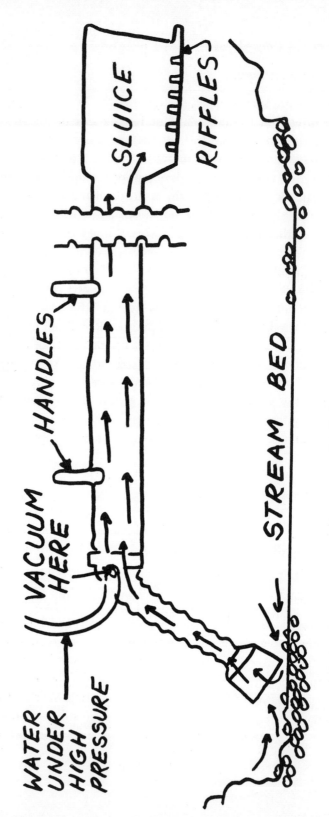

Jet dredge.

*Working the Deep Channel Placer*

To work a buried placer, it is often necessary to tunnel into the gold-bearing gravel along the slope of a hillside. This is termed drifting or drift mining and is common in hard rock mining. This tunnel or adit is used to haul out the gold-bearing gravel and to provide drainage as well, since water is often present in a drift. Sometimes the pay gravel is hoisted up through a shaft to the surface, and the adit is used only for drainage. A shaft inclined at 45 degrees is often used by present day operators as the gravel can be moved up the inclined grade with less expensive equipment and with greater safety. In the event of equipment or power failure, the miners can walk up the inclined shaft to the surface.

A source of water is necessary to wash the sand, gravel, and clay away from the gold. Enough seepage or spring water is often available in the mine for washing the gravel, if a system is provided to circulate and re-use the water, such as a series of settling basins or ponds. This system is practical and it meets ecological and debris law requirements.

## Machines Work for the Modern Miner

Horse, mule, burro, occasionally oxen, and human muscle were the principal sources of power for the miner of a hundred years ago. Even the Depression miners of the 1930's were limited in their choice of machines to assist in their search for, and recovery of, gold. But now, 45 years later, many machines developed and built for road building, dirt moving, hauling, or other types of mining are being altered and adapted to the various types of gold mining.

Following are pictures taken of a trommel gravel washing plant which operated near Nevada City, Madison County, Montana, during the summer of 1974. This is not necessarily a typical small miner operation but shows the trommel being used as a washing plant for placer gold. In this case the water supply was limited to the small, mid-summer flow from springs feeding into Brown's Gulch. This water was caught in a pond below and pumped back to the trommel. The

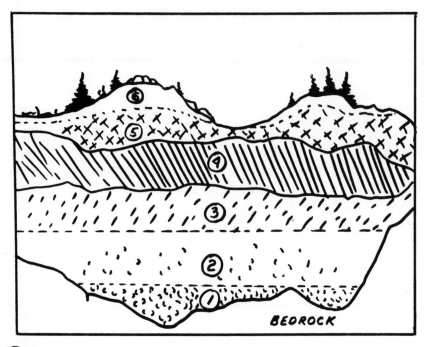

BEDROCK

① PAY GRAVEL. 1 OZ. PER YD. LOWER SECTION
② PAY GRAVEL, ½ OZ. PER YD. UPPER SECTION
③ PIPE CLAY
④ LAVA
⑤ SUB SOIL
⑥ TOP SOIL

*Deep channel placer—cross-section.*

GMG **Assay Office**

A Division of GOMIL CHEMICAL CO.
**MINERS' EXCHANGE BUILDING**

432 WEST MAIN STREET · QUINCY, CALIFORNIA 95971

PHONE: 916-283-2280
CABLE ADDRESS:
"TRANSPHERE"
QUINCY, U.S.A.

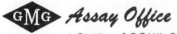

# MEMORANDUM OF ASSAY

MADE FOR _George A. Nugent_____ DATE _____ Dec. 21, ___, 196_74_

| SAMPLE NO. | PER TON OF 2000 POUNDS AVOIRDUPOIS | | | | | | | | COPPER, OR | | | LEAD, OR | | | TOTAL | |
| | GOLD | | | | SILVER | | | | | | | | | | | |
| | AT | | PER OUNCE | | AT | | PER OUNCE | | AT | | PER LB. | AT | | PER LB. | | |
| | OZS. | 100'S | $ | CTS. | OZS. | 100'S | $ | CTS. | % | $ | CTS. | % | $ | CTS. | $ | CTS. |
| 1. Outcroping | 0 | 02 | 188 | 25 | | | | | | | | | | | 3 | 77 |
| 2. Wade ? | 0 | 01 | 188 | 25 | | | | | | | | | | | 1 | 88 |
| 3. Tailings | 0 | 04 | 188 | 25 | | | | | | | | | | | 7 | 53 |
| | | | | | | | | | | | | | | | | |
| | | | | | | | | | | | | | | | | |
| | | | | | | | | | | | | | | | | |
| | | | | | | | | | | | | | | | | |
| | | | | | | | | | | | | | | | | |
| | | | | | | | | | | | | | | | | |
| | | | | | | | | | | | | | | | | |

ASSAY NO. _____

CHARGES $_7.50 Paid WEM_____

BY _____
WILLIAM E. MILLER, ASSAYER.

**CHEMISTRY** *Touches* **EVERYTHING**

*Assay report. The No. 1 sample was taken from an outcrop of a deep channel placer in Plumas County, California. It was on the top six inches of a 72 foot section of fine white quartz and was taken to identify the outcrop rather than to find values. It was a surprise to find $3.77 worth of gold in a sample that close to the top of the section.*

*The Wade Sample was from the upper section of a deep channel placer from a different mine several miles distant.*

*The Tailings (or No. 3) sample was from the waste pile of the original sluice box workings of 1923. They have been reworked twice since they were originally mined and washed. This sample is also from a Plumas County, California, deep channel placer.*

amount of water available is very important in determining the type of washing plant to be used. A system of ponds that permits sediment to settle precludes pollution of the stream and makes it possible to operate a washing plant on a small supply of water.

New and more efficient gravel washing plants are being designed and built to meet the greatly increased demand for equipment of this kind—stimulated by the five-hundred percent increase in the price of gold and the passage of the private gold ownership Act of August 1974.

*Centrifugal Gold Concentrator*

Anyone who has had experience on the farm turning the handle of an old-fashioned cream separator knows that the lighter cream is separated from the heavier skim milk by the use of centrifugal force. The milk is released from the tank above into the center of a revolving bowl containing a number of disks. If the bowl is turning at the proper speed, the comparatively heavy skim milk is thrown to the outside of the disks, collected, and flows through a spout into a bucket. The cream, being noticeably lighter, is forced to the center of the disk and bowl, and into a spout leading to another receptacle, thus completing the separation process.

In the centrifugal gold concentrator the principle of centrifugal force is used in a way similar to the cream separator. The heavier gold is separated from the sand and fine gravel as it is fed into the center of the revolving machine with a continuous stream of water. The inside of the bowl has a removeable rubber liner, contoured to form riffles, which catches and holds the gold after it is thrown to the outside of the bowl by centrifugal force when the bowl is revolving at the proper speed. The waste water, sand, and light gravel boil up over the sides of the revolving bowl and are caught by a metal shield which surrounds the bowl and then flows down into a trough leading to the discharge pipe, thus completing the separation process. If the operator wishes to

Rich, gold bearing gravel was mined from this gulch near Nevada City, Madison County, Montana, in the summer of 1974.

Brown's Gulch is part of the Alder Gulch—Virginia City, Montana area. "ALDER GULCH (includes Junction, Nevada, Fairweather, Highland, Pine Grove, Summit, and Brown's Gulch) Virginia City, 8 miles east of Alder, N.P.R.R. Discovered in 1863, has yielded over $50,000,000 by sluicing, drifting, and dredging, in Alder Gulch and other tributaries. Active in 1932, 1933, and 1934. Possibility of buried channels under basalt flows should be investigated." From Reprint of Part II of Memoir 5, Placer-Mining Possibilities in Montana, by O. S. Dingman, (1971) Bureau of Mines and Geology, Butte, Montana.

A portable trommel used as a gravel washing and gold recovery plant on Brown's Gulch near Nevada City, Madison County, Montana, July, 1974.

*The cleanup at the trommel is always a job. More than a pound of coarse gold was recovered on this one on July 1, 1974 on the gulch near Nevada City, Madison County, Montana.*

*The cleanup (gold bearing sand) from the trommel is flushed into the scoop of the front end loader and then shoveled into a centrifugal gold concentrator where much more of the sand is washed away. The concentrate can then be panned by hand to separate the last of the sand from the gold.*

*Side view of centrifugal gold concentrator. The stationary shell, supported by four legs, protects the revolving bowl inside which is driven by an electric motor or gas engine.*

*Top view of centrifugal gold concentrator. The inside of the bowl has a removable rubber liner contoured to form specially designed riffles which catch and hold the gold after it is thrown to the outside of the bowl by centrifugal force. (Photographs courtesy of Duke's Manufacturing Co., 315 Colorado River Boulevard, Reno, Nevada 89502)*

check the efficiency of the bowl he can run the discharged material into a sluice box.

The concentrator has a capacity of 4 to 5 cubic yards per hour. It can be used with material from a placer, tailings, or dump or to further concentrate the gold in the cleanup from a trommel or other type of washing plant.

# Chapter IV
# Hard Rock or Vein Gold

Surely there is a vein for the silver and a place for
gold where they find it.

*Job 28: 1*

Prospecting is a tiring job that requires at least an elementary
knowledge of geology and considerable patience. The old timers may
not have known that their practical knowledge of where and how to
find gold was called geology, but they knew it produced results. They
had their fair share of patience—but even so, many prospectors have
walked away from their diggings, discouraged, only to have someone
else come along, dig a little deeper, and strike it rich.

## Binoculars—A Prospector's Tool

In looking for faulted and mineralized areas in mountain country,
some of the old prospectors found a good pair of binoculars to be a
time- and labor-saving piece of equipment. They could then study
exposed rock over a wide area. Faulted areas, bands of colors, and
areas of broken white rock indicating quartz could be detected in this
way. Promising areas would then be checked out on foot, and the
quartz would be crushed and panned for gold.

## Faults

Faults are breaks or fractures in rock which generally extend deep

63

*Binoculars—a prospector's tool.*

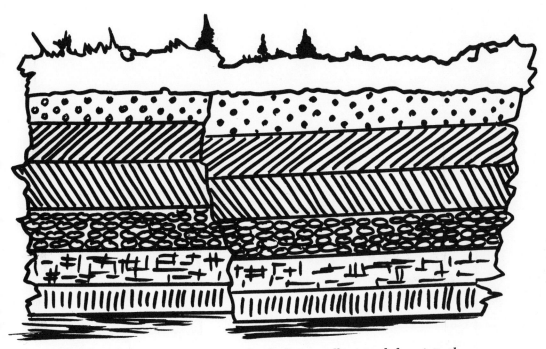

*Faults are breaks or fractures in rock which generally extend deep into the earth.*

into the earth. One side of the break will probably be lower than the other. This is called displacement and it is measured in feet and inches. The cracks and fissures resulting from the break provide a channel that permits molten minerals mixed with gas, water, and steam to rise to the surface. This material is under tremendous pressure and the cooling is often slow so crystals form. If the molten material contains quartz it may also contain gold.

Because faulted areas are highly mineralized, they display different colors which indicate valuable mineral ores, compounds, and oxides. A solid, unfaulted area of rock such as granite seldom produces valuable ores, and *never* gold.

### Red, Green, Black, and Heavy

As the old timers put it: "Look for rocks that are red, green, black, and heavy." These men had a practical knowledge of geology learned

① QUARTZ LODE WITH VEINS OF GOLD
② DECOMPOSED LODE MATERIAL WITH GOLD FRAGMENTS
③ FLOAT
④ STREAM BED
⑤ COUNTRY ROCK

RESIDUAL PLACER FROM DECOMPOSED QUARTZ LODE

*Where the nuggets come from.*

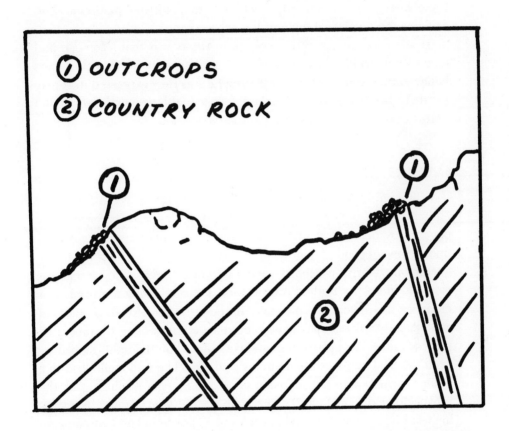

An outcrop is the edge or surface of a mineral deposit or sedimentary bed which appears on the surface.

from long experience. They had found that certain mineral compounds, including iron, manganese, copper, galena (lead sulfide), and silver were somehow associated with gold. Black sand always looked good. Red characterizes many of the iron compounds, green and blue indicate copper, manganese ores are black, galena is a light gray, and these heavy ores frequently contain both gold and platinum. This explains the geological logic of the expression.

Another old saying of the early prospectors was that "You can find iron without gold, but can't find gold without iron." These old timers believed that iron was the "mother metal," as they expressed it. It may be that they were right, because few valuable minerals are found unless iron is present in the same rock formation.

### Outcrops and Floats

Veins are often visible at the surface and are spoken of as "outcrops." A mineralized vein will often be partially or entirely covered by debris and surface material. A prospector will sometimes find a piece of good ore out loose on the surface of the ground or partly covered by debris. A piece of ore not part of a solid rock formation was called a "float" by the old prospectors. A float is like a footprint. It is evidence that a vein is somewhere in the area. Since small particles of gold, called "colors," may have been released from other disintegrated pieces of float, the prospector can take samples on the uphill side away from the float and wash them in his gold pan. The trail could lead to a hidden formation or ledge. When samples no longer produce gold, the prospector can conclude that he is above the vein. Trench from the highest point where the gold particles were last found and continue on up the hill until the mother vein is uncovered. The gold pan can be of great help to the prospector in this way.

### How Gold Was "Made"

The old prospectors discovered that sulfides, such as silver sulfide or copper sulfide, were often found in the same rock formations as gold. Quite often, very fine particles of free gold are mixed with these ores. The combination is physical, not chemical. Since it is free gold it can be recovered by smelting. They thought that iron pyrite, or fool's gold, was a young form of the sulfide ores which appeared to produce gold when processed. Consequently, they reasoned that the sulfide ores were a further stage in the formation of pure gold. What they did not realize was that the true source of the gold was the finely divided, but free, gold mixed with the sulfide ores, rather than the sulfide ores themselves.

The prospector's theory was that gold was the end product of a long series of continuous changes brought about by various forces and processes. Because gold is associated with certain minerals (particularly iron) the theory held that iron, sulfides, and quartz were necessary for gold to be brought into existence. Observation and constant checking of their theories against the contents of their gold pans convinced the old timers that this assortment of rock and minerals had to be present for gold to "make" as they called it.

This interesting bit of the old prospector's logic and theory probably wouldn't be accepted by modern day geologists, but who can prove they were wrong?

## Develop a Prospect—Follow the Lead

When the prospector was fortunate enough to make a discovery his first concern was to "stake" his claim, thus establishing his right to the mineral. Details will be given in a later section on how to stake and file on a claim. Having staked his claim, his next concern was to develop it. To develop a prospect, it will first have to be determined if the claim has values sufficient to make mining it economically feasible. Samples must be obtained and assays taken of the rock formations. As you sink a shaft or work a drift, increased gold in the assays indicates that you have what is called a "lead." One of the maxims of the old timers was to "follow your lead," and this is still good advice today. As long as the values continue to increase, the lead should guide you to a pay streak of rich ore. There should be gold in the pan after the representative ore sample is washed. When the pay streak becomes rich enough to mill or ship to a smelter, you are in business. It is just a question of getting the ore out of the ground, up to the mill, and through the processing system.

A lightweight portable drill is essential in obtaining rock samples for assay purposes. The custom built drill shown here is mounted on a two wheel trailer. It produces a pulverized, dry rock sample and delivers it to a plastic bag four feet long by four inches in diameter; representing the cuttings from ten feet of a $2\frac{3}{4}$ inch diameter hole. The profile of the formations can be studied from these bags and samples taken from any portion with a probe.

## Test Drilling and Assaying

To obtain samples of the formations on your claims for assaying, test drilling is necessary. If ground water is present in the formations it will probably be necessary to have conventional hard core samples taken by a drilling contractor. If the formations are dry, then a "dust" core, which is less expensive, may be taken. This consists of the cuttings from the bit as it drills down through the rock. These cuttings are blown and sucked up out of the hole and caught in a four-foot long by four-inch diameter plastic bag, which represents a ten foot section of the rock. The holes are drilled in a gridiron pattern, or on sites selected by a mining engineer or geologist. From the assays the location of the ore body and the value of the ore per ton can be determined. If sufficient drilling is done the total tonnage and value of the ore body can be determined.

From the cores obtained, samples are taken of only those portions that show promise, and assays are then made. Obtaining accurate assays is essential. Large operators often have their own assaying facilities, operated by company chemists. Small miners and prospectors must rely on the services of commercial assay firms. A sensible prospector will have his assaying done by a firm recognized for the high quality and accuracy of their assay reports. A potential purchaser of a mining property will seldom inspect it or consider buying unless he has some evidence of value as determined by assay reports. Also, try to have your assaying done by a firm approved by your potential purchaser. A possible buyer often requires that the assay samples be obtained by an impartial operator such as a core or test drilling company. Finally, a gold prospector should not overlook the possibility that an assay report might disclose that his claim contains minerals that might be of more value than gold!

## Develop Reserves

Another good rule is to develop a known reserve of ore—one ton in reserve for every ton mined out. Often the total effort is directed toward mining the known ore body, giving no thought to finding more

P. M. CRISMON, PRES.

# CRISMON & NICHOLS
### ASSAYERS AND CHEMISTS
440 SOUTH 500 WEST ST.

PHONE 363-7417

P.O. BOX 1708

## REPORT OF ASSAY

SALT LAKE CITY, UTAH 84110 _____ March 19, 1974 _____

WE HAVE ASSAYED YOUR ____ four ____ SAMPLES AND FIND ____ them ____ TO CONTAIN AS FOLLOWS:

William Little

| DESCRIPTION | NO. | OZS. GOLD PER TON | OZS. SILVER PER TON | PER CENT LEAD | PER CENT COPPER | PER CENT ZINC | PER CENT INSOL. | PER CENT IRON | PER CENT | VALUE OF GOLD PER TON |
|---|---|---|---|---|---|---|---|---|---|---|
| Silver Key | 1 | | None | | | | | | | |
| -"- | 2 | | 0.10 | | | | | | | |
| -"- | 3 | | Trace | | | | | | | |
| -"- | 4 | | Trace | | 0.08 | | | | | |

REMARKS:

CHARGES $ ___ 21.00 ___

CRISMON & NICHOLS

BY

_Assay report on Silver Key Claims No. 1, 2, 3, and 4, near the Salton Sea in California._

ore. If your lead plays out it can mean financial disaster if a new source of ore has not been found to replace the worked-out vein. Core drilling beyond the vein will indicate the presence of additional ore. If there is none, and the firm expects a continuous operation, new properties must be found. For more detailed information on the development of a mine prospect refer to the Montana Bureau of Mines and Geology, Miscellaneous Contribution No. 13, *Practical Guide For Prospectors in Montana.*

*Ready—Fire!*

# Chapter V
# Powder—Fire in the Hole!

Sooner or later, the old-time miner would find he could not go farther in working his vein without doing some blasting. This was when he learned to handle powder. In those days, black gunpowder (made of charcoal, saltpeter, and sulfur) was used for blasting. The black, granular powder was poured into a hole that the prospector had laboriously drilled into the hard rock with his long, chisel-like bit, heavy hammer, and strong arm. The fuse, which was in effect a small tube filled with gunpowder, was strung out as far away from the hole as required for safety. When lit, the burning powder in the tube slowly burned its way to the other end and ignited the charge. The explosion would shatter the rock so the miner could dig on for a few more feet.

## Dynamite

Although dynamite was discovered in 1866 and replaced powder for blasting purposes, the term "powder" is still used when referring to dynamite. Dynamite is an explosive mixture of glycerin, sodium or ammonium nitrate, and a filler of combustible pulp, such as wood meal. Although it is less sensitive to shock than nitroglycerin, it will still detonate if subjected to a sudden strong impact. Before you even *think* about buying dynamite, get a copy of the *Blaster's Handbook*, published by E. I. du Pont de Nemours & Co., Wilmington, Delaware. I'm not going to give you all the details you will need to know in working with dynamite, since you can get these from the handbook, but let me warn you that dynamite *is dangerous* and that this is one place where excuses don't count and where there is no second chance. You do it right the first time or you wind up doing your prospecting in the Great Mother Lode in the sky. However, there are a few things which are not in the handbook that I have learned from my own experiences; the life they save may be yours. For instance, anyone using dynamite should know that tamping with a steel rod is inviting disaster. Tamping means driving or pounding dirt or sand down into a drill hole above the charge of powder or dynamite. Almost everyone knows that steel striking a hard, flint-like rock will create a spark. A spark is one thing you *don't* want near that powder. Use wood and tamp lightly! Fresh dynamite can take about a fifty pound tap with a wooden stick.

*Old Dynamite Is Dangerous*

Dynamite left for a year or more has a tendency to form pockets or bubbles of nitroglycerin. Freezing will cause this too. Aging and freezing are the two things to watch out for with dynamite. If little white crystals have formed on the ends of the stick, handle them just as though they were pure nitro. The best thing to do is to take the sticks to the nearest open field, pour gasoline on them and then lay a cap with the proper wires attached alongside of them. Stand well back and set them off. They won't make a loud noise—just a flash of flame—and that is the end of your worry. Old dynamite is very dangerous! *DON'T try to use it! DESTROY IT!*

*Setting a Charge*

To set a charge of fresh dynamite, put as many sticks in your hole or holes as the hardness of the rock requires. Your handbook will explain this. The last stick to go into the hole has the cap, either an electrical cap or a cap and fuse. Use a wooden stick about the size of the diameter of the fuse to bore a hole about two inches deep into the dynamite, then insert the cap. Loop the wires of the fuse around the stick of dynamite, so the cap will not accidentally be pulled out if someone trips over the wires. Stand clear; hide behind a tree or a large rock, or far enough away to avoid the possibility of rocks falling on your head. Then set off the charge. A flashlight battery is enough to set off a single cap, but use one that will provide at least a six volt charge. Allow at least two volts for each additional cap, to be sure of being very safe. Be alert! A charge that does *not* detonate can still be *very* dangerous. Some men, who didn't live to tell about it, have drilled right into an area where an old charge was located, thinking it had exploded or was free from dynamite.

## A Delayed Blast

Dynamite can be deadly in peculiar ways, especially if you are careless. It almost happened to me many years ago. I was working a mine near La Porte, California; it was medium hard rock, and to put a shaft down it was necessary to use dynamite. I was down about fourteen feet when I ran into a situation that almost caused me to consider throwing away the booklet I had on how to safely handle dynamite.

I was using a fuse and caps to set it off. I put a small charge of about fourteen sticks into the hole and then crimped the cap onto the fuse with a crimper, which is a little tool similar to a pair of pliers that crimps one end of the cap around the end of the fuse. The purpose is to hold the cap onto the fuse. I lit the fuse, got out of the hole, headed for cover, and waited for the blast to come. It didn't come. I waited about five minutes more. Something told me not to go near the hole. It was about 11:15 A.M., so I decided to eat an early lunch and give it time to go out completely. Then I planned to put a fresh stick in alongside the rest with a new fuse and cap and set it off. All fifteen sticks in the hole would then detonate.

I took my time about eating. After lunch, I started back up towards the diggings from my camp. It was about 12:10 P.M., and the shaft was approximately 1000 feet from my camp. I was walking along with the fresh stick of dynamite and the fresh fuse and cap in my hand—about twenty to twenty-five feet from the hole—when the charge blew! I hit the dirt and small pieces of rock fell all around me; some of the smaller pieces hit my back and legs. It didn't hurt me, but it sure shook me up: I suddenly realized that if I had started from camp one minute sooner, I wouldn't be here to tell about it.

Needless to say, I didn't do much more that day. I decided it wasn't my day to set off dynamite. I sat down on a log and tried to figure out what had happened. I had heard old timers tell stories about such things, but I didn't think it would ever happen to me. I finally decided that this is what had happened. The crimp was too tight on the fuse. It prevented the powder or fire from reaching the cap. The outside of the fuse sat there and smouldered for nearly an hour. The fire finally reached the powder portion beyond the crimp, which caught fire, and finally detonated the charge.

*Tape the Cap*

I sat there on that log until I figured out a method to avoid any possible reoccurrence of this accident. After several minutes, I realized that the solution was to *not* crimp the cap! Then I asked myself how I could keep the fuse from coming out of the cap if it were not crimped. The thought came to me: why not *tape* the fuse onto the cap? I had some electric tape with me, so I tried it. I put the fuse into the cap and then taped it together; that way, the fuse could not come out of the cap. I tried setting it off, using just the fuse and cap without the dynamite. It worked. I then realized that this would also seal out any water—such as seepage from the drilled hole—that might make the charge fail to detonate. I tried leaving the cap and fuse to soak in water for about an hour before I lit the fuse: it still exploded. Using this method, I have never had any more trouble. I understand that the handbooks still recommend the use of crimpers to attach the fuse to the stick of dynamite, but I have a good reason to believe that my method is the better one.

*Premature Blast*

Several years later, I was working in the same general area again. At
this time, a construction crew was working on the top dam of the
Feather River Project, which is about three or four miles from La
Porte in the Little Grass Valley. While I was in town visiting a friend, I
heard several ambulances scream by, going out to the dam. About
thirty minutes later they came screaming back again, heading for the
hospital at Oroville. Still later, some of the workmen came into town
and told me that a charge of dynamite was set off accidentally by a
sudden lightning storm. The lightning bolt had struck near the
detonator box, where the wires leading to the dynamite charge were
strung out. The crew was working on the charge when the bolt struck.
Unfortunately, several men were killed and the others were seriously
hurt. As a safety precaution, one of the wires had even been
disconnected from the box and was lying on the ground—in case
someone accidentally pushed the plunger on the detonator box.

This event was a great shock to me, as it was to most of the people
in that small town. I had trouble sleeping that night. I tried to figure
out how this terrible accident had happened and how such a thing
could be avoided in the future. To some extent this was a personal
problem for me; I had been using electric caps to set off my dynamite
for several years. I found them to be more convenient than a cap and
fuse, as I could decide to the split second when the charge was to be
set off—a simple matter of pushing the handle of the detonator.

*Lead Seal*

After a time, I realized that there was a simple solution to the
problem. Everyone who handles dynamite and electric caps should
know that the caps come from the manufacturer with a small lead seal
attached across the two wires that run to the powder part of the cap.
This lead seal shorts out the wires, making it impossible for the cap to
detonate. Many good powder men don't seem to realize that this lead
seal should *never* be broken until the wires are all strung out and the

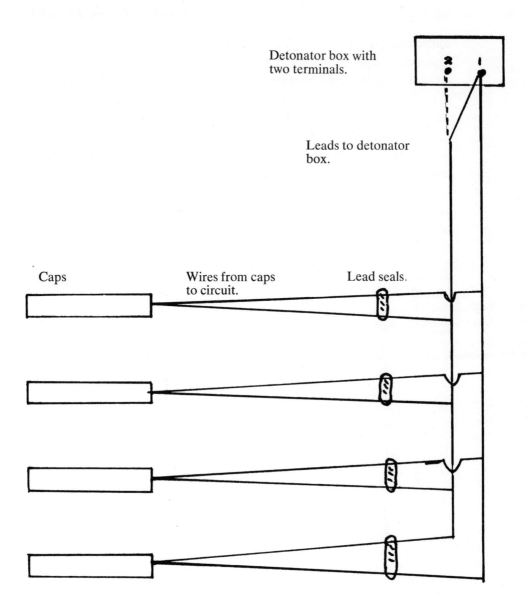

Detonator box with two terminals.

Leads to detonator box.

Caps

Wires from caps to circuit.

Lead seals.

*How to prevent premature detonation. This sketch shows the circuit from the detonator box to four electric caps. The lead seal across the two wires of each cap creates a dead short which prevents premature detonation. These seals should not be broken until you are ready to fire the charge.*

*The leads to the detonator box should both be attached to the No. 1 terminal until ready for firing, then one wire should be removed and attached to the No. 2 terminal. This creates an open circuit ready for firing.*

entire charge ready to go. The too-early breaking of this seal has been a primary cause in many seemingly mysterious dynamite accidents. In stringing out the wire from the detonator box to the site of the charge, it is common practice—as a safety measure—to drop one of the two wires to the ground instead of connecting it to the terminal on the box, just as was done by the powder man at the dam accident site. Of course, the wire would be attached to the terminal just before the operator was ready to set off the blast.

*Short the Circuit*

In analyzing what had happened in the lightning accident, I concluded that the lightning bolt could not have completed the circuit to the electric cap if the operator had temporarily placed both wires on one terminal, instead of dropping one wire on the ground. This would have created a dead short, capable of blocking any flow of electricity to the caps.

In practice, the only absolutely safe procedure is to keep the two wires connected to *one* terminal of the detonator box, and to keep the seals on the electric caps *unbroken* until all the wires are strung out and the charge is ready to go. Then, when the operator is sure that all the men are clear, he can send one man to the charge to remove the seals from the electric caps. When this man is clear, the operator can take the second wire from the first terminal and place it on the second terminal, thus opening the circuit to the electric caps. The entire system is now ready. All that remains is for the operator to shout "Fire in the hole," and push down the plunger. If this very simple routine is followed, the possibility of an accident is reduced to practically zero, even if lightning does strike nearby as it did in the dam accident.

*Transmitter Can't Cause Blast*

Most people traveling by automobile through areas where road construction and blasting are going on have noticed signs stating

"Turn off Transmitter—Blasting." Such signs would be entirely unnecessary if the procedures outlined above were followed; the use of these signs indicates that either the man in charge doesn't understand the basic principles explained above, or that he is compelled to put up the signs by some law or regulation that is a result of ignorance. There is no way that a radio transmitter can cause a premature detonation when the electric cap seals are unbroken and the lead wires are shorted out by placing them both on a single terminal of the detonator box.

If you are a weekend gold panner, steer clear of dynamite: you don't need it. If you are serious about hard rock mining, *be careful!* If you follow the precautions above you shouldn't get in serious trouble with dynamite.

### Two Component Explosive

Recently, an entirely new form of explosive has come into the picture. It utilizes two individual components—a liquid and a powder—which can be stored or transported as non-explosives until they are mixed on the site. This is unique because neither of the components alone is explosive, but when mixed together they become as powerful an explosive as dynamite. The explosive is mixed and used in two forms: as a stick and in a pouch. When the dry powder in the stick or plastic pouch is armed by pouring in the liquid it can be readily ignited with either electric or fuse caps. This new explosive was developed by the Atlas Powder Company. Additional information may be obtained by writing this company at P. O. Box 2354, Wilmington, Delaware 19899.

I was informed by another source that the idea for this type of explosive was the result of a tremendous accidental explosion which occurred on the Gulf Coast several years ago. A ship loaded with ammonium nitrate fertilizer blew up at the dock, devastating the area and killing a large number of people. Subsequent investigation disclosed that fuel oil from the storage tanks for the diesel engines of the ship may have leaked into the cargo of fertilizer. Experiments with this type of mixture disclosed it to be highly explosive when ignited.

# Chapter VI
# How to Stake a Mining Claim

Before discussing how to stake a claim and what that involves, it should be clearly understood that we are only talking about Federally owned land. These areas are known as public domain or public lands and are administered by the U. S. Forest Service under the Department of Agriculture, or by the Bureau of Land Management under the Department of the Interior. Under Federal regulations, it is only on these public lands that you may prospect for minerals and stake a claim. All lands in the United States not owned by the Federal Government are privately owned and are not open to prospecting unless arrangements are made with the owners.

## What Is a Mining Claim?

Briefly, an unpatented mining claim is an area of land on which an individual has obtained a right to extract and remove minerals by the act of valid location under the mining laws of 1872, but where full title has not been acquired from the United States Government. This is known as a possessory right, and is maintained by the performance of certain prescribed annual work and recording requirements. This right may be sold, inherited, or taxed according to state law.

Mining claims were officially brought into existence by an Act of

85

Congress in 1872, called "An Act to Promote the Development of Mining Resources of the United States." The complete law is found in the United States Code, Title 30, Sections 2k-54. The regulations can be found in the Code of Federal Regulations (CFR), Title 43, parts 3400-3600, which are available from any Bureau of Land Management office.

## Mining Districts

Prior to the enactment of these basic mining laws of 1872, the regulation of mining activities was done through "Mining Districts." After the discovery of gold became known and people arrived in numbers it became evident that some form of government had to be organized to control mining activity and to protect the rights of the individual miners. Meetings were called by the miners who owned claims staked on a certain creek or land area. At these open meetings, which were conducted by a leader selected by those present, rules were set down to establish and protect the property rights of the miners. Each land area was defined and named as a particular mining district. As other problems related to keeping the peace, establishing order, and administering these rules arose, mining districts functioned as government until the organization of the Territories. A central place was designated where the claim boundaries and their dates of delineation were recorded. This was generally a designated building in the largest settlement or town of the mining district which then became the seat of government and was the prototype for our present day counties and courthouses.

When formal government was established in the territories and states, the original need for mining districts disappeared. Today, they are of historical interest only, although the name given to each district still identifies specific areas within the political subdivision of the county and state. The forms used to record mining claims, called "notices of location," still require the name of the mining district in which the claim is located. In addition, the notice requires a description of the location of the mining site (called the "discovery") in relation to some permanent and natural object within a stated county, range, township, and section. This ties the old mining district designations to the currently established system of land survey and the legal description of land areas, as well as to the county subdivision of government. Location notices of claims are recorded in the county courthouse as a permanent, public record.

### Discovery of a Valuable Mineral Deposit

To be valid, a mining claim may be staked and held only after the discovery of a valuable mineral deposit. The courts have established what is called the "prudent man" rule to determine or define what constitutes a valuable mineral deposit. The discovery is declared valuable if a person of ordinary common sense (prudence) would be justified in a further expenditure of his labor and means to develop this mine, with a reasonable prospect of success. The basis for the prudent man rule and the authority for all federal mining law comes from the CFR, Title 43, Code mentioned above.

### State Mining Laws

Each state has individual mining laws, so there are various differences in the requirements from state to state. These laws parallel the Federal Code, however, and do not conflict with it. The laws of the state in which you wish to stake your claims may be secured from your State Bureau of Mines or your local public library.

### Staking the Lode Claim

Federal law requires that a claim be marked distinctly enough so that it can be readily identified. Each state has expanded this general requirement to include detailed directions for marking claim boundaries. The minimum requirements are listed as follows:

1. Set a stake at the point of discovery and one on each corner of the claim. A center stake may be required on each side. Some states require the name of the claim to be painted on each stake, with a designation on each stake regarding center, or direction; i.e., Northwest corner, Southwest corner, Center, etc.

2. The location notice, giving the name of the locators, the name of the claim, whether lode or placer, the description of the location,

*Staking the claim.*

any minerals claimed, and meeting any other requirements made by the state, must be placed at the point of discovery. This should be placed in a waterproof container, plainly visible beside the discovery monument or stake.

3. An exact copy of the location notice must be filed at the county Clerk and Recorder's office in the county in which the claim is located.

4. A lode claim is limited in size by statute to a maximum of 1500 feet in length and 300 feet on both sides of the discovery lode or 600 feet total width. The longer sides are parallel to the course of the vein. The location is described by metes and bounds, a term meaning that the direction in degrees and distance in feet is measured from point to point and return to the place or point of beginning.

5. If the location description of either a lode or placer claim is not tied in with the legal subdivision system, it must provide the distance and direction as accurately as possible from the discovery point to some well-known reference point, such as a point on a steel bridge, a fork in a stream, a road intersection, or some other known landmark within the county.

Diagram of the Sweetgrass Lode claim.

| | |
|---|---|
| R | Reference point (corner of steel bridge) |
| D | Discovery of valuable mineral |
| XY | Course (or direction) of vein |
| AB | Northeast boundary line of claim, 1500 feet long, parallel to course of vein |
| CS | Southwest boundary line of claim, 1500 feet long, parallel to course of vein |
| BS | Southeast end of claim, 600 feet wide |
| AC | Northwest end of claim, 600 feet wide |
| NB and MA | Distance from center of vein to northeast boundary line—300 feet. |
| NS and MC | Distance from center of vein to northwest boundary line—300 feet. |

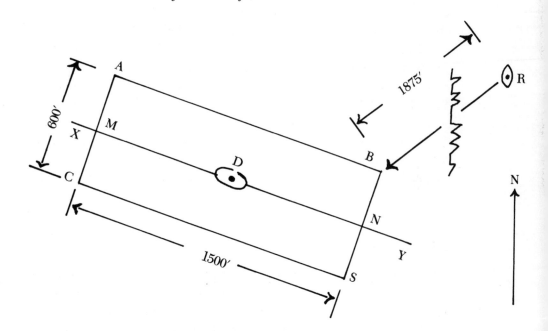

## Metes and Bounds Land Descriptions

Metes are measures of length in feet and bounds are boundaries. A metes and bounds description locates the land area with reference to permanent objects or readily distinguished landmarks in the vicinity of the land area to be described.

Following is a description of how the various points on the diagram are determined:

1. The line XY is determined by the point of discovery "D" and the course or direction of the vein.

2. Points M and N are determined by measuring 750 feet each way from point D along the course of the vein or line XY and including point D, thus establishing the length of the claim at 1500 feet, the maximum length by statute for a lode claim.

3. From point N measure 300 feet on a perpendicular or a 90 degree angle from XY. This establishes point "B."

4. In like fashion, points S, A, and C are established, thus determining the two boundary lines AB and CS parallel to the course of the vein XY.

5. Points A and C, and B and S, when connected form lines which comprises the two ends, 600 feet in width, parallel to each other—the maximum width by statute for a lode claim.

6. The next step is to find some point of reference—in other words, some permanent and recognizable place to tie to. We select the northwest corner of a steel bridge over Ruby River, 4 miles south of Twin Bridges, Montana, as this point of reference which we designate as point "R."

7. With the aid of a compass, we determine the angle from point R to point B and measure the distance from R to B, which we find to be 1875 feet. We now know the location of point B with reference to point R.

### Describing Your Claim

For the purpose of preparing a description to identify the location of your claim on the location notice, you can estimate the approximate direction from your reference point on the corner of steel bridge as simply "Southwest." By stepping off the distance you find it is 625 steps of three feet each, giving 1875 feet to the nearest corner "B." Now, you can state the location of your claim on the location notice as follows:

From the southwest corner of the Roscoe Bridge on the Ruby River, Madison County, Montana; approximately 1875 feet in a southwesterly

*Surveying the claim.*

RECEIVED FOR RECORD
FEB 27 1974

At Request of
_____ Min. Past _____ o'clock, _____ M.

Book 1974, Page 22968
Recorded in Official Records
of Riverside County, California

W.O. Boesch Recorder

FEES $ _____

22968

(SPACE ABOVE THIS LINE FOR RECORDER'S USE)

# NOTICE OF LOCATION

## LODE

TO WHOM IT MAY CONCERN:

Please take notice that the undersigned locator(s): are citizens of the United States or have declared their intention to become a citizen; hereby claim the following described Lode; and post the following notice of location.

1. The name of this lode or claim is the **SILVER KEY # 5** _____ lode mining claim; and is located in **OROCOTIA M.D.**, County of **RIVERSIDE**, State of California.

2. The name, current mailing address or current residence address of the locator(s) are given at the bottom of this notice.

3. The number of linear feet claimed in length along the course of the vein each way from the point of discovery is **750** feet in a **N.W.** direction. (Not to exceed 1500 feet)

feet in a **S.E.** direction, and **750** feet in a _____ direction. (Not to exceed 1500 feet)

4. The width on each side of the center of the claim is **300** feet. (Not to exceed 300 feet)

5. The general course of the vein or lode is _____

6. The date of the posting of the original of this Notice, which is the date of location, is **FEB 22, 1974**.

7. This claim, described by reference to some natural object or permanent monument as will identify the claim located, is as follows:

**SEC 20 T6S, R11E SBM. 50 OF HWY 195**
**JOINS SILVER KEY #3 ON NE SIDE (WHICH IS THE**

LOCATORS

| NAME | STREET ADDRESS | CITY, STATE and ZIP CODE |
|---|---|---|
| 1. W. A. LITTLE | 203 E. 4ᵀᴴ ST. | BOLDER, MONT. 59633 |
| 2. VERNE H. BALLANTYNE | 217 W. KOCH ST. | BOZEMAN, MONT. 59715 |
| 3. | | |
| 4. | | |

## STATEMENT OF THE MARKING OF THE BOUNDARIES
## AND OF PERFORMANCE OF DISCOVERY WORK

NOTICE IS HEREBY GIVEN by the undersigned locator(s) that in accordance with the requirements of the California Public Resources Code:

1. The above notice of location is a true copy of said notice; and is hereby incorporated by reference herein and made a part hereof.

2. The locator(s), within the following time, as required by law, have defined the boundaries of this claim by erecting at each corner of the claim and at the center of each end line, or the nearest assessable points thereto, a conspicuous and substantial monument; and each corner monument so erected bears or contains markings sufficient to appropriately designate the corner of the mining claim to which it pertains and the name of the claim. The date of marking is: _____ ; and the description of monuments are:

_____

3. The United States survey within which all or any part of the claim is located is: Township _____,
Range _____ and Meridian _____.

4. The locator(s) have performed the following location work: _____

_____

DATED: _____, 19 ____        LOCATOR(S) _William H. Little_

_____

# DO NOT RECORD

1. This form should not be used where a placer location notice, tunnel site location notice, or mill site location notice is required.

2. Section 2316, Public Resources Code, reads as follows:

"(a) A wooden post or stone structure not less than 3½ inches in diameter or a metal post not less than two inches in diameter, projecting at least three feet above the ground and set at least one foot in the ground, or a mound of stone at least three feet in height above the ground, is prima facie a conspicuous and substantial monument as referred to in this chapter.

(b) Where by reason of precipitous ground, it is impractical or dangerous to place a monument in its true position, a witness monument may be erected as near thereto as the nature of the ground will permit and marked so as to identify the true position.

(c) Where by reason of working the claim, it is impractical or dangerous to maintain a monument in its true position, a witness monument shall be erected as near thereto as the nature of the ground will permit and marked so as to identify the true position."

3. Within 90 days after the posting of a Notice of Location, the locator(s) " . . . shall record in the office of the county recorder of the county in which such claim is situated a true copy of the notice together with a statement by the locator of the markings of the boundaries, and of the performance of the location work, as well as the character of each, which statement also shall include the township, range and meridian of the United States survey within which all or any part of the claim is located."

Section 2312, Public Resources Code.

———————

This standard form covers most usual problems in the field indicated. Before you sign, read it, fill in all blanks, and make changes proper to your transaction. Consult a lawyer if you doubt the form's fitness for your purpose.

*This is a standard NOTICE OF LOCATION (lode) form. Various printing establishments make these forms and they are sold in stationery stores wherever there is a need for them.*

*The lower part of this form was not completed as the discovery work was not yet done at the time of filing. Assay reports from samples taken from an outcrop at the time of filing were negative so the discovery work was never done and the claim lapsed. The form begins on page 92.*

direction to point "B," the northwest corner of this claim; from point B 1500 feet northwest, parallel to the trend of the discovery vein "D," to point "A," thence 90 degrees southwest 600 feet to point "C," thence southeast 1500 feet, parallel to the trend line, to point "S," thence northeast 600 feet to point "B," the point of beginning.

This improvised description, while not entirely accurate, is sufficient and legal as far as staking your claim is concerned. (Note: This is not an actual location.)

### Surveying the Claim

Neither the Federal nor the state laws require that mining claim location descriptions be established by an official surveyor. However, if you find that you have valuable minerals and start to mine or prepare your claims for sale, it is a wise policy to have a survey made by a licensed surveyor and to have this more accurate description recorded in the county Clerk and Recorder's office.

A description prepared by a qualified surveyor of your claim would read like this:

Beginning at point "R" (the southwest corner of the Roscoe Bridge on Ruby River), thence south 79 degrees—15 minutes on a true bearing a distance of 1875 feet to point "B," the true point of beginning, thence south 18 degrees—45 minutes west, a distance of 600 feet to point "S," thence north 71 degrees—15 minutes west a distance of 1500 feet to point "C," thence north 18 degrees—45 minutes east 600 feet to point "A," thence south 71 degrees—15 minutes east a distance of 1500 feet to point "B" the true point of beginning, lying in SW¼SE¼ Sec. 30 T14N, R23W.

### Discovery Work

Part of the requirement to establish a valid claim is the performance of certain discovery work. State mining laws vary somewhat in their requirements, but a ten foot deep pit or shaft is generally required. Drilling a twenty-five foot test hole can usually be substituted for the pit or shaft. Ordinarily, a certain period of time is allowed from the time the claim is staked or filed until the required discovery work must be completed. In some states, this period is ninety days. You must secure additional information to be sure that you are complying with the laws in the state where you are staking your claim.

### Placer Claims

Placer claims are limited in size to 20 acres per claim per locator. An association of two or more persons, up to a total of eight, may locate 20 acres per locator, making a total or maximum of 160 acres. Where practicable, placer claims are located by legal subdivision. A legal subdivision is a section or a part of a section lying in a designated township, range, and principal meridian, for instance, S½SE¼ Sec. 14, T6N, R9E, MPR. On unsurveyed land or where not practical because of the terrain, placer claims may be located by metes and bounds.

### Assessment Work

A minimum of $100 worth of work must be performed each year on, or for the benefit of, your claim in order to maintain your right to this claim. This work ordinarily consists of extending the depth of the shaft, the length of the tunnel, or doing other work on, or for, the benefit of the claim. This work must be done sometime during the mining year (September 1st to September 1st) and must be completed on or before noon on September 1st of each year. A statement that this work has been performed must be recorded in the county recorder's office within a specified period after September 1st. Again, state requirements are different. Be sure you have the right information for your state. This work requirement is known as annual assessment work.

### Patenting the Claim

The rights of the owner of a mineral claim are limited to possession for the purpose of developing and extracting discovered valuable minerals. To obtain the additional rights included in full ownership, a patent must be obtained. This can be quite expensive, as there are many legal requirements that must be satisfied before a patent can be issued. Among these requirements are:

1. Proving a valuable discovery.
2. Survey by cadastral engineer.
3. File a plat of claim.
4. Post 60-day notice on claim.
5. Apply for patent—$25.
6. Not less than $500 work.
7. Full description of vein.

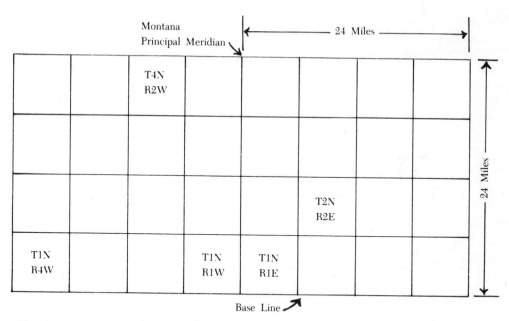

*U.S. Governmental Survey System—The Tract. The largest squares measure 24 miles on each side and are called checks or tracts. Each tract is further divided into 16 squares called townships whose four boundaries each measure six miles. The columns of townships running north and south are called ranges and are numbered according to their distance from the principal meridian (the main north-south survey line).*

U.S. Governmental Survey System—The Township. A township is six miles square or 36 square miles. Each square mile is designated as a section which is equivalent to 640 acres.

The rows of townships running east and west are numbered according to their distance from the baseline (the main east-west survey line).

Sections within a township are numbered from the northeast corner, following a back-and-forth course until the last section in the southeast corner is reached. For purposes of land description, sections are commonly divided into half-sections containing 320 acres, or quarter-sections containing 160 acres. Land descriptions are made by referring to a particular quarter of a particular section located within a particular township, county, and state.

U.S. Governmental Survey System—The Section. The 40 acres in the southeast corner of the Section 2 above, located in Township 5 North and Range 6 East would be legally described thus:

SE ¼ of the SE ¼ Section 2, Township 5 North and Range 6 East of the Montana Principal Meridian, County of Gallatin, State of Montana.

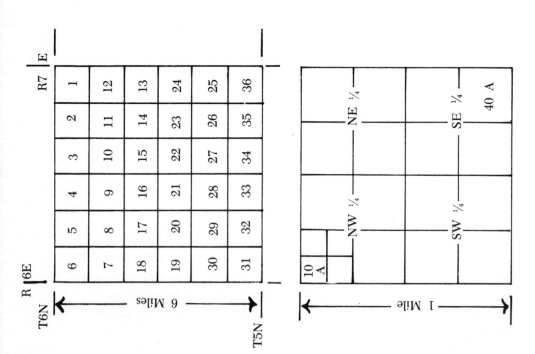

8. If placer, prove is placer.
9. Prove possessory rights.
10. Other supporting papers.

From this partial list of requirements, you can see that obtaining a patent is not a simple procedure. Anyone intending to file an application for a patent of their mining claims should contact a Bureau of Land Management or Forest Service office for complete information.

If you are discouraged by this list of requirements, you may be consoled by the fact that a patent to your mining claim is not necessary for carrying on mining activity, as long as the legal requirements to maintain your right as a claimant are performed. These requirements are 1. being able to show the discovery of a valuable mineral, and 2. performing a minimum of $100 of annual assessment work and recording this in the Clerk and Recorder's office in the courthouse of the county in which the claim is located.

For more complete information on how to obtain a patent to a mining claim you are referred to a small pamphlet available from the Bureau of Land Management, U. S. Department of the Interior, entitled *Patenting a Mining Claim on Federal Lands—Basic Procedure*. This is a free publication available from any of the twelve Land Offices of the Bureau of Land Management. For Montana and North and South Dakota the address is Montana Land Office, 316 North 26th St., Billings, Montana 59101. For Arizona it is Arizona Land Office, Federal Building, Phoenix, Arizona 85025.

## The Small Hard Rock Miner

The hard rock miner can be "in business" on practically any scale, from small to large, although in general rock mining requires more original investment and more knowledge of mining than is required of the placer miner. It is safe to say that the ore from every mine has its own individual characteristics and requires a method of handling and processing specifically related to those characteristics. To go any further into the methods of handling or processing ore is beyond the scope of this book.

However, the following two pictures, and the accompanying description, are included to give a general idea of one rather typical small hard rock miner's operation from the handling and processing point of view. This miner was able to secure a second hand crusher at a reasonable price, and since he was a welder and mechanic, built the

*This millsite in Josephine County, Oregon, is equipped with an old-style crusher on the left and a high speed dry mill on the right. The ore is crushed to nut size in the crusher, then put in the dry mill where it is reduced to a powder. The dry mill is custom built and similar to a huge fan or water wheel with four vanes or paddles. The ore is mined from a nearby outcrop.*

dry mill himself. The shaker table was also a used piece of equipment.

This is definitely a "poor boy" operation as far as the equipment is concerned. However, the most modern and expensive mining and handling machine does not guarantee, or even imply, success in a small mine operation. The knowledge and experience of the miner or the guiding hand and intelligence of his engineer or advisor is the critical element in a successful small mine operation.

From the dry mill, the powdered ore is blown into a cyclone where it is watered down with a spray, then passed onto a shaker table where the waste rock floats off and the heavy concentrate of minerals and metals is shaken into pans at the foot of the table. The slope of the table, the amount of water allowed to flow onto it, and the amount of vibration are all important factors in this operation.

*President Ford signs the private ownership of gold bill—August 14, 1974.*

# Chapter VII
# Ownership—Marketing—
# Federal Regulation

Several other factors are important for the amateur prospector and miner to know and are discussed below.

## Private Ownership of Gold

During the period of economic stress and depression of the early 1930's the GOLD RESERVE ACT OF 1934 was passed by the Congress. This Act arbitrarily required all persons in the United States to exchange their privately owned gold and bullion for so-called lawful (paper) money. By this Act it became illegal for a citizen of the United States to own gold coin or bullion. This law, together with related legislation, regulations, and the licensing of gold buyers coupled with the increasingly unrealistic and arbitrarily fixed price of gold in effect destroyed the market for gold for the small miner.

Prior to the passage of this Act it was common practice for the gold miners to ship their gold to the nearest United States Mint where it would be graded, refined, and purchased or minted into coin or bullion for the owners. This free coinage of gold by the United States Treasury provided a ready market for the producers of gold so long as the arbitrarily fixed price was in balance with other segments of the economy. There was no question concerning the private ownership of gold at that time. Ownership, in the usual sense, was assumed and taken for granted.

On August 14, 1974, President Ford signed Public Law 93-373 of the 93rd Congress which repealed the GOLD RESERVE ACT OF 1934 and made it legal for any person in the United States to buy, sell, hold, or otherwise deal in gold either in this Country or abroad as of December 31, 1974. Thus, a right which was taken from United States citizens over forty years ago was restored. This historic action officially legalized the private ownership of gold.

This Bill, S.2665, in its final form, was attached to an amendment to a Bill relating to participation of the United States in the International Development Association. This is a subject unrelated to the ownership of gold and illustrates the peculiar manner in which the Congress sometimes makes known its will.

Because of its historical significance an exact copy of the entire amendment is reproduced below.

o   o   o

Public Law 93-373
93rd Congress, S. 2665
August 14, 1974

AN ACT—88 STAT. 445

To provide for increased participation by the United States in the International Development Association and to permit United States citizens to purchase, hold, sell, or otherwise deal with gold in the United States or abroad.

*Be it enacted by the Senate and House of Representatives of the United States of America in Congress assembled,* That the International Development Association Act (22 U.S.C. 284 et seq.) is amended by adding at the end thereof the following new section:

"Sec. 14. (a) The United States Government is hereby authorized to agree to the Association four annual installments of $375,000,000 each as the United States contribution to the Fourth Replenishment of the Resources of the Association.

"(b) In order to pay for the United States contribution, there is hereby authorized to be appropriated without fiscal year limitation four annual installments of $375,000,000 each for payment by the Secretary of the Treasury."

Sec. 2. Subsections 3 (b) and (c) of Public Law 93-110 (87 Stat. 352) are repealed and in lieu thereof add the following:

"(b) No provision of any law in effect on the date of enactment of this Act, and no rule, regulation, or order in effect on the date subsections (a) and (b) become effective may be construed to prohibit any person from purchasing, holding, selling, or otherwise dealing with gold in the United States or abroad.

"(c) The provisions of subsections (a) and (b) of this section shall take effect either on December 31, 1974, or at any time prior to such date that the President finds and reports to Congress that international monetary reform shall have proceeded to the point where elimination of regulations on private ownership of gold will not adversely affect the United States' international monetary position."

Sec. 3. The International Development Association Act (22 U.S.C. 284 et seq.) is amended by inserting at the end thereof the following:

"Sec. 15. The United States Governor is authorized and directed to vote against any loan or other utilization of the funds of the Association for the benefit of any country which develops any nuclear explosive device, unless the country is or becomes a State Party to the Treaty on the Non-Proliferation of Nuclear Weapons (21 UST 483)."

Approved August 14, 1974.

This legislation was largely the result of the untiring efforts and dedication of Senator James A. McClure and his aide Al Timothy, from Idaho, in the Senate, and Congressman Philip M. Crane of Illinois, aided by fifty co-sponsors of the Bill, in the House of Representatives.

## Market for Gold

With the passage of the private ownership of gold Act, quoted above, gold became a commodity in the same sense that copper, silver, platinum, wheat, corn, lumber, cotton, or any other product bought, sold, and used in domestic and international trade is a commodity. Upon the passage of the Act, the major commodity markets or exchanges of the nation offered to buy or sell gold futures contracts. There is no restriction on the purchase or sale of gold—it may be sold to anyone who wishes to buy. As time goes on and the gold mines of the country again become productive, new marketing channels will be opened up on a domestic level.

In the past, the price of gold has been determined at certain central world markets, such as Zurich, Switzerland, and London, England.

These markets will continue to be a major factor in the pricing of gold. Unlike other commodities, the historic role of gold as a standard of value to support domestic currencies and regulate international settlement of trade balances places an unknown factor on the pricing of gold in these world markets. Political considerations of an international origin modify or distort the action of a free market.

A very special and profitable market exists for gold nuggets in their natural state. These nuggets, also called specimens, may be sold for several times the regular world market price for gold. Unusual pieces of hard rock gold, such as quartz containing veins of gold, are also in demand as specimens and will bring premium prices from collectors and hobbyists far in excess of the value of their gold content.

### Federal Regulations

Prospectors and miners should be reminded that the Congress of the United States has charged the Secretary of Agriculture (and Secretary of the Interior) with the responsibility for administering certain laws it has passed with regard to the use of the surface of the National Forest System lands. To comply, the Secretary of Agriculture authorized the Forest Service to formulate and enforce new rules and regulations relating to the preservation and use of the surface of Forest Service lands. These rules were published in the Federal Register, Volume 39, Number 168, August 28, 1970, together with a resume of the hearings held prior to their final writing, and became effective on September 1st, 1974. Copies of this issue of the Federal Register may be obtained from any Forest Service District Office.

These rules are of special interest to prospectors and miners as they set up requirements in certain situations relating to their activity. As long as the prospector does his prospecting on foot or with a horse or pack animal as the old timers did, or confines his motorized vehicle to established roads and doesn't tear through the forest with a bulldozer, he will have no problem. However, if the prospector believes it necessary to strike off in a four wheel rig where there has previously been no road, or wishes to use a bulldozer to make a trench for discovery or assessment work, he will find it necessary to file a notice of intent with the District Ranger. If the Ranger decides that the

*Hey—Mister Prospector!*

operation may cause significant disturbances to the surface, the operator will be required to submit an operational plan.

By knowing the regulations and planning ahead, the smart prospector will probably find that he can carry on the work necessary for his discovery and subsequent assessment work without being in conflict with any existing regulations. When, and if, his development work requires significant disturbance of the surface, he will find the Forest Service ready to cooperate with him in developing a plan that will permit development of the mine and protect the environment as well. The prospector should contact the Forest Service a reasonable time ahead of his proposed activity to permit them to arrange time to discuss and work out the problem with him.

In addition to the considerations above, the prospector who has made a discovery and filed his claims must be prepared to defend his discovery at any time; that is, he should be able to show that he has discovered a valuable mineral in a sufficient enough quantity to warrant its further development. This is a basic requirement of the mining law of 1872 and even though the Forest Service does not have the authority to enforce the requirement, they can and do check new claims for the validity of the discovery. If a claim shows no evidence of value they will notify the Bureau of Land Management, which does have the authority to contest and cancel the claim.

In recent years, some individuals have abused the privileges granted under the mining law of 1872 by staking claims where there were no discoveries of valuable minerals in order to secure a free summer home or camp. This practice has made it necessary for the Federal government to supervise discovery and claim activity more closely.

# Conclusion

Now that we have told you the secrets of the old '49ers they are no longer secrets in one sense of the word. *Webster* defines a secret as information "hidden from others" or "revealed to none or to a few" so perhaps we can still say they are secrets since we have revealed them to only a few; to you and a few others. This is still only a small number, considering how many prospectors and miners are going to be around looking for gold in the next few years. In any event, we have enjoyed passing on some of the old timers' secrets to you and hope they may be helpful.

Bill Little, my partner, is the one who really contributed most of the secrets. He spent over 25 years up in the High Sierra gold country in northern California after he got out of the Air Corps in 1945. He knew most of the old timers who remained up in that country, and he learned a lot from his own experience. You can't beat experience. When you get to checking out those outcrops or tiltin' a pan on some likely creek you'll find that you will soon be your own teacher, doing what comes next, and naturally. That's experience!

After Bill told me some of the old timers' theories I decided to check some of them out in the books. I mean the geology books. I read several of them and found that those old timers were pretty well up on their geology—even though they probably couldn't even spell the word.

One day, Bill and I got to talking, and I said "Bill, why don't we write a little six or ten page pamphlet on how to find gold? We'll mimeograph it and maybe sell a few." He said, "It's all right with me,

109

*Many of the old and forgotten mines will again be opened and made to produce their gold.*

if you do the writing." Well, that started the ball rolling: I found that six or ten pages weren't enough to hold all the old timers' secrets that Bill came up with, so we had to make it larger. One thing led to another, until the pamphlet turned into a booklet, then into a book and that is what you have in your hand.

Bill told me to be sure to wish you a lot of luck in your prospecting—he said you would need it. I'll join with him in the wish, so we both say, "GOOD PROSPECTING, PARDNER," and if you strike it rich from anything you learned out of this book we would sure be glad to hear about it. Just drop us a line in care of our publishers and they will pass it on.

Verne H. Ballantyne

# Glossary

Definition of Mining Terms

(Courtesy Montana Bureau of Mines and Geology)

ADIT     A horizontal gallery or opening driven from the surface of the ground to the ore body. The term "tunnel" is frequently used in place of adit, but technically a tunnel is open to the surface on both ends.

ALLUVIAL     Generally pertains to loose gravel, soil, or mud which has been deposited by water.

ANALYSIS     A separation of compound substances by chemical means.

ASSAY     The determination of the valuable minerals in a sample. A wet assay is determined by the use of chemicals. A fire assay is determined by both chemicals and fire. Gold and silver are usually assayed by fire.

BEDROCK     Any solid rock underlying gold-bearing gravels.

BLACK SAND     Grains of heavy, dark minerals such as magnetite, limenite, chromite, etc., found in streams which commonly collect in sluice boxes and which may carry gold and platinum.

CHUTE     An opening in the ground where ore is allowed to pass from one level to another. It is the structure built to load cars from a stope.

CLAIM      A land area claimed by a prospector and marked out by stakes.

COLOR      A term referring to small grains or flakes of gold.

CONTOUR      Lines connecting points of equal elevation on a contour map.

DIP      The maximum angle of inclination downward that a vein or bed makes with a horizontal plane.

DYNAMITE      Dynamite is an explosive mixture of glycerin, sodium or ammonium nitrate, and a filler of combustible pulp such as a wood meal.

ELECTRIC CAP      A small metallic cap containing fulminating powder which is detonated by an electric current.

EXPOSURE      Any part of a vein or rock outcrop that can easily be seen.

FAULT      A fracture in the earth, with displacement of one side of the fracture with respect to the other.

FISSURE      An opening or crack in the rock. A fissure vein is a fissure filled with mineral matter.

FLOAT      The loose and scattered pieces of ore which have been broken off from an outcrop.

FOOTWALL      The bottom or lower enclosing wall of a vein.

FUSE      A tube or cord filled or impregnated with combustible matter for igniting an explosive charge after a predetermined interval, as in blasting.

HANGING WALL      The top or upper enclosing wall of a vein.

HEAD FRAME      A structure erected over a shaft to support the sheave wheel for hoisting purposes.

HEADING    Any part of a mine where work is under way. Usually confined to development workings only.

HIGH GRADING    Stealing of high grade ore or nuggets from the workings of a hard rock or placer mine by employees or others.

IGNEOUS ROCK    Rock formed from molten lava.

LATERAL    A horizontal mine working. A drift in the footwall of a vein is often called a lateral.

LEASE    A contract by which one conveys real estate for life, for a term of years, or at will, usually for a specified rent or royalty.

LESSEE    A person who obtains a lease on mining land.

LESSOR    The grantor of a lease.

LEVEL    All the connected horizontal mine openings at a certain elevation.

LOCATING    The marking of the boundaries and staking of a mining claim.

LODE    Refers to a tabular deposit between definite walls.

MILLING ORE    Ore that must be concentrated at or near the mine before it is shipped.

MUCK    A common term for any broken ore or underground waste.

MUCKER    A shoveler, or one who handles muck.

NITRO    Short for nitroglycerin, which is any nitrate of glycerol, a colorless, heavy, oily, explosive liquid used in making dynamite.

NUGGET    A piece of gold of any shape or size larger than a flake, usually rounded by stream and water action.

OPTION    This is the right to purchase at a stated price.

ORE    A mineral aggregate of sufficient value to be mined at a profit.

ORE BODY    The part of a vein that carries ore. Generally, all parts of a vein are not ore. Ore shoot has the same meaning.

OUTCROP    The edge or surface of a mineral deposit or sedimentary bed which appears on the surface.

OVERBURDEN    The valueless material overlaying the pay zone in a placer deposit or the waste or valueless material of a solid outcrop.

OXIDE    A compound of a metal and oxygen.

PATENT    A written title to land granted by the government after meeting certain obligations. A mining claim can be patented after $500 worth of work has been done and other requirements met.

PLACER    Alluvial deposit of valuable mineral-bearing gravel.

PLANE    An even surface. A horizontal plane is a flat, even, level surface.

POWDER    A miner's term for dynamite or other explosive.

RAISE    An excavation of restricted cross-section, driven upwards either vertically or at an angle from a level in the mine.

RAKE    The trend of the ore body within the vein.

RIFFLE    Grooves, channels, slats, or wire screens in a sluice box or rocker to catch gold or other valuable minerals.

SLIP    A small fault.

SLUICE BOX    A trough paved with riffles through which gravel and wash from placer mining operations pass so that gold and other valuable minerals can be caught and saved.

SPECIFIC GRAVITY    The ratio of the weight of any substance to the weight of an equal volume of water.

STOPE    Any excavation underground used to remove the ore.

STRIKE    The bearing of a horizontal line in the plane of a vein, bed, or fault in respect to the cardinal points of the compass.

STRIPPING     Removal of the overburden from a placer deposit or the barren outcrop from an ore deposit.

STULL     A timber used to support loose rocks or slabs. It may also be used to support a platform in a working area.

TREND     The general direction or bearing of a vein, fault, or rock outcrop.

VALUE     Refers to the mineral substance searched for. In the case of gold the term is synonymous with color.

VEIN     A well-defined, tabular, mineralized zone which may or may not have valuable ore bodies.

WALL     The waste or country rock on either side of a vein.

WASTE     Barren rock or mineralized material which does not have enough value to be classified as an ore.

WORKING FACE     Any portion of the mine where work is under way, such as the face of a drift or the face of a raise.

# Bibliography

## Books

*Bacon and Beans From a Gold Pan.* Jesse L. Coffee and George Hoeper. Doubleday & Co., 1972, Garden City, N.Y.

A hardcover book which is in print and sells for $5.95. It is an account of the personal experiences of the author, Jesse L. Coffee, and his wife Dot, during the Depression years of the 1930's in the Mother Lode placer areas in California, where they made their living working the gravel bars in the streams of the High Sierras. It is interesting reading and gives many practical ideas on placer mining.

*Blaster's Handbook.* E. I. du Pont de Nemours & Co., Wilmington, Delaware.

A practical and complete manual giving instructions for setting up a charge, both electric and fuse; quantities of dynamite for different hardnesses of rock; safety precautions; and many details the powder man and amateur miner or prospector should know.

*Handbook For Prospectors and Operators For Small Mines.* W. M. Von Bernewitz, revised by Harry C. Chellson.

This large, 531-page, hardcover handbook is out of print, but may be available in your library. It contains much valuable mining information, is somewhat technical, and some of the material and pictures are out of date. It was originally copyrighted in 1926 but has been revised and the copyright was renewed in 1963.

SPECIAL NOTE: An up-to-date, fifth edition of the *Handbook for Prospectors*, based on the Von Bernewitz book and edited by Richard M. Pearl, is now available from McGraw-Hill Book Co., 1221 Avenue of the

117

Americas, New York, N.Y. 10020. It is 596 pages, 5½" × 8", and sells for $14.50.

*Placer Examination—Principles and Practices.* John H. Wells, Mining Engineer, Bureau of Land Management. Available from the Superintendent of Documents, Washington, D.C. 20402.

A hardcover book containing excellent information on the use and construction of rockers, sluice boxes, and mechanical gold washing machines, with numerous pictures and sketches. It sells for $1.50.

*Practical Guide For Prospectors and Small Mine Operators in Montana.* Koehler S. Stout. Montana Bureau of Mines and Geology, Butte, Montana.

A 103-page softcover booklet with six pages of drawings and sketches, which contains excellent information on prospecting, estimating, developing, mining methods, placer mining, Montana Mining Laws, Revised Codes of 1947; a table of chemical elements and a list of references; and lists of assayers, consulting engineering firms, machinery suppliers, and miscellaneous mining and equipment suppliers. It is in print and a bargain at $1.00 per copy.

*Principal Gold Producing Districts of the United States.* A. H. Koschmann and M. H. Bergendahl, 1968. Available from the Superintendent of Documents, Washington, D.C. 20402.

A 283-page book which gives a thorough, descriptive listing of all the gold mining districts in the United States. It gives production data and geologic information of value, and sells for $4.75.

*Prospecting and Operating Small Gold Placers.* William F. Boericke. John Wiley & Sons, Inc., N.Y.

A hardcover book, 7½" × 5", 143 pages, originally copyrighted in 1933, and now in its second edition and 11th printing. Some of the data on equipment was updated in 1960 at its last printing. This excellent book covers all phases of placer mining and has proven its acceptance by its numerous printings. It is in print and sells for $5.95.

*Shallow Diggins—Tales From Montana's Ghost Towns.* Jean Davis. Caxton Printers, Ltd., 1962, Caldwell, Idaho 93754.

A collection of historical narratives of people and events in the various mining camps and settlements in early Montana. This book is very interesting and informative, is now out of print, but is available in libraries.

*Small Fortunes in Penny Gold Stocks.* Norman Lamb. SRT Corporation, P.O. Box 148, Vallejo, California 94590.

How to make money by buying penny gold stock is the basic theme of this very interesting and informative book. The author is apparently knowledgeable on the practical aspects of mining and geology as well as the skill and timing of the market so that the book gives much valuable information on the economic and business end of gold and silver mining. It is a 1974 publication and is available at $10.00 per copy.

*Treasure Hunter's Guide (How and Where to Find It).* Robert I. Nesmith and John A. Potter. Arco Publishing Co., Inc., 219 Park Avenue South, New York, N.Y. 10003.

Discusses all types of treasure hunting, and the section on skin and scuba diving is full of information and encouragement for the underwater crevice miner who wishes to use floating dredge equipment. This book sells for $5.95.

*The World of Gold.* Timothy Green. Walker Publishing Company, 1973, New York, N.Y.

A general treatise on gold: the history of its use, mining, and political and monetary considerations.

## Other Publications

*American Gold News*, P.O. Box 457, Ione, California 95640.

A monthly newspaper-type publication containing many articles and news of interest to the gold prospector. Deals entirely with gold-related information and political developments. Published monthly; $5.00 per year.

*California Geology*, P.O. Box 2980, Sacramento, California 95812.

A monthly publication issued by the California Division of Mines and Geology; subscription price $2.00. It contains excellent articles on many phases of mining of interest to the prospector and small miner.

*California Mining Journal*, 2539 Mission Street, P.O. Drawer 628, Santa Cruz, California 95061.

The main trade journal of the mining industry in California and adjoining states. It has timely and comprehensive articles on various aspects of mining. Published monthly; subscription price $5.00 per year.

Denver Equipment Co., Division of Joy Manufacturing Co., 1499 Seventeenth Street, Denver, Colorado 80217.

This firm sells mechanical gold-saving equipment, and provides various gold evaluation bulletins and articles in addition to sales information. Available upon request, without cost.

*Engineering and Mining Journal*, 2539 Mission Street, Santa Cruz, California 95061.

The leading trade journal of the mining industry in the United States, containing articles of national interest to miners. Published monthly.

*Gold Districts of California—Bulletin No. 193.* Published by the California Division of Mines and Geology, Ferry Building, San Francisco, California 94111.

Principal features of each gold-bearing district in the state are described in some detail.

*Gold Placers of California—Bulletin No. 92.* Charles Scott Haley. Published by California State Mining Bureau, Ferry Building, San Francisco, California 94111.

Brief descriptions of the location and extent of the principal gold placers in California, including maps. This is a 165-page publication by Charles Scott Haley, a prominent private consulting mining engineer, written in 1923 for the California State Mining Bureau after two years of intensive field work. It is now out of print and available only in libraries, but is a very valuable source of data on the principal placer mining areas in California.

*How to Mine and Prospect for Placer Gold*, Bureau of Mines Information Circular No. 8517. J. M. West. Available from the Superintendent of Documents, Washington, D.C. 20402.

This Bureau of Mines publication includes a brief history of placer mining in the United States. It discusses where to look for placers by area and state, and discusses pollution problems related to placering. 43 pages, extensive bibliography.

*Laws and Regulations Covering Mineral Rights in Arizona.* Published by the Department of Mineral Resources, State of Arizona, Phoenix, Arizona.

A booklet containing the laws and regulations relating to mining in Arizona. Essential for anyone expecting to stake a claim in Arizona or interested in mining in that State. Available at $1.00 per copy.

*Legal Guide For California Prospectors and Miners*, Special Publication 40. Published by the California Division of Mines and Geology, Ferry Building, San Francisco, California 94111.

An invaluable source of information on all legal phases of mining in California. It is the most popular item published by the Division of Mines and Geology, and is in its 13th edition and fourth printing. Contains 134 pages, and is available for $1.00 per copy.

*Locating Gold*, United Prospectors, Inc., 5665 Park Crest Drive, San Jose, California 95118.

A membership, hobby-type organization. The membership fee of $6.00 includes a magazine, which is published every two months.

*Patenting a Mining Claim on Federal Lands*. Bureau of Land Management, U.S. Department of the Interior, Washington, D.C.

Provides basic information on how to file an application for a patent and describes the administrative procedures involved. Revised 1970; free.

*Pay Dirt*, P.O. Drawer 48, Bisbee, Arizona 85603.

The official publication of the Arizona Small Mine Operator's Association. Contains many articles on various phases of mining in Arizona, including political news related to mining. Published monthly; $5.00 per year.

*Placer Mining for Gold in California—Bulletin No. 135*. Charles Volney Averill. Published by the California Division of Mines and Geology, 1923, Ferry Building, San Francisco, California 94111.

A valuable bulletin on placer mining in California. It is now out of print and available only in libraries.

*Staking a Mining Claim on Federal Lands*. Bureau of Land Management, U.S. Department of the Interior, Washington, D.C.

Provides, in question and answer style, fundamental facts on where and how to stake a mining claim. Discusses size and shape of lode and placer claims and other important information. May be obtained from any Bureau of Land Management Office or from the Superintendent of Documents, U.S.

Government Printing Office, Washington, D.C. 20402. Revised 1970; priced at 15 cents.

*Western Prospector and Miner*, P.O. Box 146, Tombstone, Arizona 85638.
   A newspaper-type publication containing many articles and news items of interest to the gold miner and prospector. Published monthly; subscription price $5.00 per year.

# Index

# About the Author

Verne H. Ballantyne has been prospecting for gold for twenty-five years. His interest in gold mining began in the early 1900's when a few of the old-timers, eager to share their secrets, were still around, actively seeking gold. He spent many evenings around the campfire listening as these men re-lived their experiences on the trail of gold and later worked with them on their claims. Mr. Ballantyne is currently a real estate broker in Montana.